Bioengineering Liver Transplantation

Bioengineering Liver Transplantation

Special Issue Editors

Luc J.W. Van der Laan
Bart Spee
Monique M. A. Verstegen

MDPI • Basel • Beijing • Wuhan • Barcelona • Belgrade

MDPI

Special Issue Editors

Luc J.W. Van der Laan
Erasmus University Medical Center
The Netherlands

Bart Spee
Utrecht University
The Netherlands

Monique M. A. Verstegen Erasmus
University Medical Center The
Netherlands

Editorial Office
MDPI
St. Alban-Anlage 66
4052 Basel, Switzerland

This is a reprint of articles from the Special Issue published online in the open access journal *Bioengineering* (ISSN 2306-5354) in 2019 (available at: https://www.mdpi.com/journal/bioengineering/special_issues/liver_trans)

For citation purposes, cite each article independently as indicated on the article page online and as indicated below:

LastName, A.A.; LastName, B.B.; LastName, C.C. Article Title. *Journal Name* **Year**, *Article Number*, Page Range.

ISBN 978-3-03921-744-1 (Pbk)
ISBN 978-3-03921-745-8 (PDF)

Cover image courtesy of Monique M. A. Verstegen.

Contents

About the Special Issue Editors

Luc J.W. Van der Laan is based at Erasmus MC—University Medical Center Rotterdam, the Netherlands, where he is Professor in Liver Regenerative Medicine and Head of the Laboratory of Experimental Transplantation and Intestinal Surgery (LETIS). The research conducted at tis center has a strongly translational character and aims to provide the basic and applied knowledge for improving the outcome of patients with end-stage liver failure or liver cancer. Novel regenerative strategies involve human liver and bile duct organoids, bioscaffolds and organ perfusion technologies.

Bart Spee is Assistant Professor at Utrecht University. Here, he combines stem cell and biofabrication technologies in order to create in vitro models that mimic liver function in great detail. This, in turn, would replace animal testing for, for instance, toxicity testing. Moreover, he is interested in using the same technology to create entire organs via whole organ engineering, which would alleviate issues from liver donor shortage. Bart Spee is also coordinator of the annual Regenerative Medicine and Advanced In Vitro Models summer school in Utrecht.

Monique M. A. Verstegen is trained as a biologist and has a background in adult stem cell biology. She is currently Assistant Professor at the Erasmus MC—University Medical Center Rotterdam, the Netherlands, at the Department of Surgery. She focuses on research directed at liver and bile duct regeneration and liver disease modeling using organ-on-a-chip technology. She has great expertise in using organoid technology to model liver and primary liver tumors.

Preface to "Bioengineering Liver Transplantation"

Liver transplantation is the only effective treatment for end-stage liver disease, but the shortage of organ donors severely limits the number of patients that benefit from this therapy. An understanding and critical evaluation of emerging concepts in regenerative medicine, biomaterials, and stem cell biology is therefore essential for the bioengineering of human liver grafts. With this, novel liver transplantation procedures and personalized alternative treatments of end-stage liver disease become available and can be applied. This Special Issue of Bioengineering covers the search for alternatives, which include the optimization of organ preservation by oxygen persufflation to improve poor quality liver grafts and make them suitable for transplantation. In addition, one paper presents 3D printing technology to manufacture scaffolds for patient-specific liver tissue or bile ducts as a novel approach. In this paper, a novel 3D-printed stent infused with collagen, mesenchymal stromal cells (MSC), and cholangiocytes is validated for its use in personalized biliary procedures. In addition to this specific bioprinting approach for bile ducts, different types of bioprinting technology for liver tissue engineering are reviewed, such as laser-based bioprinting, inkjet bioprinting, and extrusion bioprinting. Bioengineered liver tissue is not only developed as an alternative for donor organs, but also as a promising tool to use in drug testing, toxicological studies, and in disease modeling. Next to the organ-typical cells, supportive materials such as hydrogels are used. The utilization of these hydrogels in liver tissue engineering, as well as more bioengineering alternatives for liver regenerative medicine are reviewed in this issue. Finally, an excellent example of the implementation of genetically engineered liver cells in a canine model for copper toxicosis is described. We would like to thank all contributing authors, reviewers, and the editorial office for assisting in the establishment of this Special Issue on Bioengineering Liver Transplantation.

Luc J.W. Van der Laan, Bart Spee, Monique M. A. Verstegen
Special Issue Editors

bioengineering

MDPI

Editorial

Bioengineering Liver Transplantation

Monique M.A. Verstegen [1,*], Bart Spee [2] and Luc J.W. van der Laan [1]

[1] Department of Surgery, Erasmus MC-University Medical Center Rotterdam, 3000 CA Rotterdam,
The Netherlands; l.vanderlaan@erasmusmc.nl

[2] Department of Clinical Sciences of Companion Animals, Faculty of Veterinary Medicine, Utrecht University,
3584 CT Utrecht, The Netherlands; B.Spee@uu.nl

*   Correspondence: m.verstegen@erasmusmc.nl

Received: 9 October 2019; Accepted: 14 October 2019; Published: 16 October 2019

Since the first in-man liver transplantation was performed by Starzl et al. [1] in the early 1960s, many patients have successfully undergone organ transplantation. To date, over 30,000 liver transplantations are performed annually worldwide, which is estimated at less than 10% of the global need [2]. Transplantation is the only curative treatment for end-stage liver disease with different etiology. These include fat, alcohol- or viral hepatitis-related cirrhosis, liver cancer, inherited and metabolic diseases, and acute liver failure [3]. With demand increasing, alternatives for donor organs are urgently needed. In addition to optimizing donor/recipient selection procedures [4] and allocation systems to assure maximal utility of donor organs and equity to recipients [5], investing in innovating bioengineering technology remains important [6].

In the last decade, dynamic (hypo- or normothermic) machine perfusion protocols were introduced to preserve donor livers as opposed to static cold storage which, until then, was the standard [7–9]. In addition to preservation, normothermic machine perfusion can be used to assess the quality of the graft by monitoring liver function. Moreover, machine perfusion has the potential to improve grafts of marginal quality by applying treatment using (stem) cells, drugs, and compounds, for instance, to reduce steatosis, infection, or ischemia/reperfusion injury. Oxygenation of the liver graft seems to be a critical factor during machine perfusion. New developments are the use of gaseous oxygen, is based on venous systemic oxygen persufflation (OPAL) during static cold storage to improve hepatic energy homeostasis to prime the liver for the critical warm reperfusion which is known to be responsible for most of the ischemia/reperfusion injury [10]. As reported by Gallinat et al., this technique was shown to be safe and cost-effective, and demonstrated preliminary beneficial effects on clinical outcomes in a single-center randomized controlled clinical trial.

Potential alternatives for liver transplantation may include allogenic hepatocyte transplantation, in which the transplanted cells engraft the recipient's liver and restore liver function [11]. So far, the success of hepatocyte transplantation depends on the disease type and level of cell replacement that is required to restore liver function. Although promising for some diseases, the long-term efficacy remains limited. Alternatively, genetics, gene-editing, and matrix-based culture methods of diseased hepatocytes could be employed to cure the patient's own cells ex vivo as a personalized method for diseases with a known genetic mutation, as reviewed by Kruitwagen et al. [12]. Due to the (challenging) need of large numbers of hepatocytes coming from suboptimal livers that are unsuited for transplantation, cell function is often hampered [13]. To overcome this, (induced) pluripotent stem cells (iPSCs) or adult liver stem cells, such as cultured as liver organoids [14], might be a good source to use in cell transplantation [15]. Both cell types can be expanded to gain high numbers and could be initiated from the patient's own tissue, preventing the need to treat the patient with lifelong immunosuppressive drugs.

Next to functional cells, the actual organ scaffold is of key importance for the maintenance and potential engineering of functional liver tissue. If cells lack this spatiotemporal control of physical and biochemical cues, they cannot properly function and likely dedifferentiate [16]. A supporting scaffold that provides physical and biochemical characteristics, such as stiffness and matrix composition, is therefore essential [17]. Such scaffolds can be obtained from native liver tissue by decellularization [18–20]. These decellularized liver scaffolds can either be used for recellularization by infusion with cells [21] or processed into a biological hydrogels for clinical-grade expansion of stem cells or organoids [22]. Scaffolds can also be made from synthetical hydrogels, as reviewed by Ye et al. [23–25], each having their own (dis)advantages. Combining natural with synthetic hydrogels might increase the overall performance of tissue-engineered liver grafts [26]. Hepatocyte viability and proliferation might even be improved by adding even more components to the scaffold, such as special conduction polymers and gelatin, chitosan, and hyaluronan [27]. Kryou et al. reviewed novel opportunities for printing technology to create 3-dimensional (3D) scaffolds that can be used as a basis for functional liver tissue [28]. Printing technology to customize biliary stents are summarized by Boyer et al. [29]. By infusing these stents with collagen, mesenchymal stromal cells, and patient-derived cholangiocytes [29], personalized tissue engineering is nearing clinical application.

The diverse contributions in this Special Issue on Bioengineering Liver Transplantation provides an overview of the novel and exciting opportunities in the field of liver regenerative medicine, biomaterials, and stem cell research that can be applied in future transplantations and personalized treatments of end-stage liver disease.

Conflicts of Interest: The authors declare no conflict of interest

References

1. Starzl, T.E.; Marchioro, T.L.; Vonkaulla, K.N.; Hermann, G.; Brittain, R.S.; Waddell, W.R. Homotransplantation of the liver in humans. *Surg Gynecol Obs.* **1963**, *117*, 659–676.
2. Who-ONT-Collaboration. Global observatory on donation and transplantation (godt) data. Available online: http://www.transplant-observatory.org/data-charts-and-tables/ (accessed on 11 October 2019).
3. Fox, A.N.; Brown, R.S. Is the patient a candidate for liver transplantation? *Clin. Liver Dis.* **2012**, *16*, 435–448. [CrossRef] [PubMed]
4. Flores, A.; Asrani, S.K. The donor risk index: A decade of experience. *Liver Transplant.* **2017**, *23*, 1216–1225. [CrossRef] [PubMed]
5. Tschuor, C.; Ferrarese, A.; Kümmerli, C.; Dutkowski, P.; Burra, P.; Clavien, P.-A.; Lendoire, J.; Imventarza, O.; Crawford, M.; Andraus, W.; et al. Allocation of liver grafts worldwide is there a best system? *J. Hepatol.* **2019**, *71*, 707–718. [CrossRef] [PubMed]
6. Bizzaro, D.; Russo, F.P.; Burra, P. New perspectives in liver transplantation: From regeneration to bioengineering. *Bioengineering* **2019**, *6*, 81. [CrossRef] [PubMed]
7. Quillin Iii, R.C.; Guarrera, J.V. "In 10 years" of debate: Pro—machine perfusion for liver preservation will be universal. *Liver Transplant.* **2016**, *22*, 25–28. [CrossRef] [PubMed]
8. Halazun, K.J.; Quillin, R.C.; Rosenblatt, R.; Bongu, A.; Griesemer, A.D.; Kato, T.; Smith, C.; Michelassi, F.; Guarrera, J.V.; Samstein, B.; et al. Expanding the margins: High volume utilization of marginal liver grafts among >2000 liver transplants at a single institution. *Ann. Surg.* **2017**, *266*, 441–449. [CrossRef]
9. Ravikumar, R.; Jassem, W.; Mergental, H.; Heaton, N.; Mirza, D.; Perera, M.T.P.R.; Quaglia, A.; Holroyd, D.; Vogel, T.; Coussios, C.C.; et al. Liver transplantation after ex vivo normothermic machine preservation: A phase 1 (first-in-man) clinical trial. *Am. J. Transplant.* **2016**, *16*, 1779–1787. [CrossRef]
10. Gallinat, A.; Hoyer, D.P.; Sotiropoulos, G.; Treckmann, J.; Benkoe, T.; Belker, J.; Saner, F.; Paul, A.; Minor, T. Oxygen persufflation in liver transplantation results of a randomized controlled trial. *Bioengineering* **2019**, *6*, 35. [CrossRef]
11. Gramignoli, R.; Vosough, M.; Kannisto, K.; Srinivasan, R.C.; Strom, S.C. Clinical hepatocyte transplantation: Practical limits and possible solutions. *Eur. Surg. Res.* **2015**, *54*, 162–177. [CrossRef]
12. Kruitwagen, H.S.; Fieten, H.; Penning, L.C. Towards bioengineered liver stem cell transplantation studies in a preclinical dog model for inherited copper toxicosis. *Bioengineering* **2019**, *6*, 88. [CrossRef] [PubMed]
13. Ibars, E.P.; Cortes, M.; Tolosa, L.; Gómez-Lechón, M.J.; López, S.; Castell, J.V.; Mir, J. Hepatocyte transplantation program: Lessons learned and future strategies. *World J. Gastroenterol.* **2016**, *22*, 874–886. [CrossRef] [PubMed]

14. Huch, M.; Gehart, H.; van Boxtel, R.; Hamer, K.; Blokzijl, F.; Verstegen, M.M.A.; Ellis, E.; van Wenum, M.; Fuchs, S.A.; de Ligt, J.; et al. Long-term culture of genome-stable bipotent stem cells from adult human liver. *Cell* **2015**, *160*, 299–312. [CrossRef] [PubMed]

15. Hirabayashi, M.; Goto, T.; Hochi, S. Pluripotent stem cell-derived organogenesis in the rat model system. *Transgenic Res.* **2019**, *28*, 287–297. [CrossRef] [PubMed]

16. Lee, J.S.; Shin, J.; Park, H.-M.; Kim, Y.-G.; Kim, B.-G.; Oh, J.-W.; Cho, S.-W. Liver extracellular matrix providing dual functions of two-dimensional substrate coating and three-dimensional injectable hydrogel platform for liver tissue engineering. *Biomacromolecules* **2014**, *15*, 206–218. [CrossRef] [PubMed]

17. Arriazu, E.; Ruiz de Galarreta, M.; Cubero, F.J.; Varela-Rey, M.; Pérez de Obanos, M.P.; Leung, T.M.; Lopategi, A.; Benedicto, A.; Abraham-Enachescu, I.; Nieto, N. Extracellular matrix and liver disease. *Antioxid Redox Signal* **2014**, *21*, 1078–1097. [CrossRef]

18. Verstegen, M.M.A.; Willemse, J.; van den Hoek, S.; Kremers, G.J.; Luider, T.M.; van Huizen, N.; Willemssen, F.E.; Metselaar, H.J.; IJzermans, J.N.M.; van der Laan, L.J.W.; et al. Decellularization of whole human liver grafts using controlled perfusion for transplantable organ bioscaffolds. *Stem Cells Develpment* **2017**, *26*, 1304–1315. [CrossRef]

19. Bühler, N.E.M.; Schulze-Osthoff, K.; Königsrainer, A.; Schenk, M. Controlled processing of a full-sized porcine liver to a decellularized matrix in 24 h. *J. Biosci. Bioeng.* **2015**, *119*, 609–613. [CrossRef]

20. Wu, Q.; Bao, J.; Zhou, Y.-J.; Wang, Y.-J.; Du, Z.-G.; Shi, Y.-J.; Li, L.; Bu, H. Optimizing perfusion-decellularization methods of porcine livers for clinical-scale whole-organ bioengineering. *BioMed Res. Int.* **2015**, *2015*, 785474. [CrossRef]

21. Willemse, J.; Lieshout, R.; van der Laan, L.J.W.; Verstegen, M.M.A. From organoids to organs: Bioengineering liver grafts from hepatic stem cells and matrix. *Best Pract. Res. Clin. Gastroenterol.* **2017**, *31*, 151–159. [CrossRef]

22. Ijima, H.; Nakamura, S.; Bual, R.P.; Yoshida, K. Liver-specific extracellular matrix hydrogel promotes liver-specific functions of hepatocytes in vitro and survival of transplanted hepatocytes in vivo. *J. Biosci. Bioeng.* **2019**, *128*, 365–372. [CrossRef] [PubMed]

23. Ye, S.; Boeter, J.W.B.; Penning, L.C.; Spee, B.; Schneeberger, K. Hydrogels for liver tissue engineering. *Bioengineering* **2019**, *6*, 59. [CrossRef] [PubMed]

24. Mattei, G.; Di Patria, V.; Tirella, A.; Alaimo, A.; Elia, G.; Corti, A.; Paolicchi, A.; Ahluwalia, A. Mechanostructure and composition of highly reproducible decellularized liver matrices. *Acta Biomater.* **2014**, *10*, 875–882. [CrossRef] [PubMed]

25. Mattei, G.; Magliaro, C.; Pirone, A.; Ahluwalia, A. Bioinspired liver scaffold design criteria. *Organogenesis* **2018**, *14*, 129–146. [CrossRef] [PubMed]

26. Huang, H.; Oizumi, S.; Kojima, N.; Niino, T.; Sakai, Y. Avidin–biotin binding-based cell seeding and perfusion culture of liver-derived cells in a porous scaffold with a three-dimensional interconnected flow-channel network. *Biomaterials* **2007**, *28*, 3815–3823. [CrossRef]

27. Tahmasbi Rad, A.; Ali, N.; Kotturi, H.S.R.; Yazdimamaghani, M.; Smay, J.; Vashaee, D.; Tayebi, L. Conducting scaffolds for liver tissue engineering. *J. Biomed. Mater. Res. Part A* **2014**, *102*, 4169–4181. [CrossRef]

28. Kryou, C.; Leva, V.; Chatzipetrou, M.; Zergioti, I. Digital bioprinting for organ (or liver) transplantation. *Bioengineering* **2019**, *6*.

29. Boyer, C.J.; Boktor, M.; Samant, H.; White, L.A.; Wang, Y.; Ballard, D.H.; Huebert, R.C.; Woerner, J.E.; Ghali, G.E.; Alexander, J.S. 3d printing for bio-synthetic biliary stents. *Bioengineering* **2019**, *6*, 16. [CrossRef]

bioengineering

MDPI

Review

New Perspectives in Liver Transplantation: From Regeneration to Bioengineering

Debora Bizzaro, Francesco Paolo Russo and Patrizia Burra *

Department of Surgery, Oncology and Gastroenterology, Gastroenterology/Multivisceral Transplant Section, University/Hospital Padua, 35128 Padua, Italy; debora.bizzaro@gmail.com (D.B.); francescopaolo.russo@unipd.it (F.P.R.)
* Correspondence: burra@unipd.it; Tel.: +39-049-8212892

Received: 3 July 2019; Accepted: 10 September 2019; Published: 11 September 2019

Abstract: Advanced liver diseases have very high morbidity and mortality due to associated complications, and liver transplantation represents the only current therapeutic option. However, due to worldwide donor shortages, new alternative approaches are mandatory for such patients. Regenerative medicine could be the more appropriate answer to this need. Advances in knowledge of physiology of liver regeneration, stem cells, and 3D scaffolds for tissue engineering have accelerated the race towards efficient therapies for liver failure. In this review, we propose an update on liver regeneration, cell-based regenerative medicine and bioengineering alternatives to liver transplantation.

Keywords: liver regeneration; end-stage liver diseases; regenerative medicine; liver tissue bioengineering; liver bioreactors

1. Introduction

Acute and chronic liver diseases are leading causes of morbidity and mortality worldwide, accounting for about 1–2 million deaths annually [1]. The most prominent causes of acute liver failure include viral hepatitis, alcoholic liver disease, non-alcoholic fatty liver disease (NAFLD), drug-induced liver injury, and autoimmune liver disease [2,3].

Liver transplantation is the ultimate solution in the treatment of such severe liver dysfunctions. Despite the relatively high postoperative survival rate, there are many problems to be solved, however, including a chronic donor shortage, immune rejection, and ethical issues. Therefore, cell-based regenerative therapies and novel technologies such as liver-on-chip [4] and bioprinted liver [5] are expected to be the next-generation therapies.

These innovative approaches are all based on the extraordinary capacity of the liver to regenerate. For this reason, increasing our knowledge of liver regeneration mechanisms could bring significant benefits in the treatment of liver failure and may help patients needing large liver resections or transplantation.

In the present review, we propose an update on liver regeneration, cell-based regenerative medicine approaches, and bioengineering alternatives to liver transplantation, along with futuristic approaches to overcome hurdles in liver tissue engineering.

2. Liver Regeneration

2.1. Overview of Liver Development

Hepatocytes and cholangiocytes, the two main liver cell types, are derived from the endoderm germ layer. This layer develops from the anterior primitive streak during gastrulation and is identifiable 6 h post-fertilization in zebrafish, by embryonic day 7.5 in mouse, and in the third week of human gestation [6]. The endodermal germ layer forms a primitive gut tube in which the regions of foregut, midgut, and hindgut can be identified. Fate mapping studies in mouse indicate that the embryonic liver originates from the ventral foregut endoderm by embryonic day 8.0 of gestation (e8.0) [6]. The hepatic endoderm cells, identified as hepatoblasts by e9.5, delaminate from the epithelium and invade the adjacent mesenchyme of the septum transversum to form the liver bud [7,8]. The hepatoblasts are bipotential cells and, during maturation, those residing next to the portal veins become biliary epithelial cells, while the majority of hepatoblasts in the parenchyma differentiate into hepatocytes [9]. During this process, the liver acquires its characteristic tissue architecture [10]. The balance in the numbers of hepatocytes and cholangiocytes from hepatoblasts is strictly controlled by integrated signaling and transcriptional pathways. The differentiation of hepatoblasts towards a biliary epithelial phenotype is controlled by the Jagged–Notch pathway [11,12], while hepatocyte differentiation is promoted by hepatocyte growth factor (HGF) and oncostatin M (OSM) [13]. Gradually, as the liver's development proceeds towards the final stages of maturation, which begins by e13 and continues until several weeks after birth, there is a marked decline in the number of hepatoblasts [14]. However, some of the bipotent progenitor cells do not differentiate and gradually stop proliferating, establishing the pool of hepatic progenitor cells (HPCs) [15].

2.2. Homeostasis and First Line of Response to Injury

The liver has a variety of functions fundamental to homeostasis, including bile secretion, metabolism, serum proteins production, glycogen storage, and drug detoxification. Since the Ancient Greek era with the famous "Prometheus" myth, the liver has been known to have a strong intrinsic regenerative ability in vivo. Thanks to a number of evolutionary protections, this physiological process of liver regeneration allows the recovery from even substantial hepatic damage caused by toxins or viral infections [16].

Hepatic regeneration, enabling the liver to continue to perform its complex functions despite a significant injury, is crucial to the survival of mammals and is therefore evolutionarily conserved and pathways leading to its completion are essentially redundant [17].

After the loss of tissue or an injury, the liver responds with fine-tuned pathways of regeneration via the activation of a wide array of signaling and transcriptional factors. As such, after surgical partial hepatectomy, the liver's mass and function are restored within a week [16]. In epithelial tissues with a high turnover, such as the intestines and the skin, cellular renewal and tissue homeostasis is performed by a pool of stem cells. In the liver, however, the turnover is low with a mature hepatocyte having a life expectancy of about 200 days [18]. The general assumption, until recently, was that all mature hepatocytes were able to divide to ensure normal liver homeostasis [19,20]. Now the prevailing theory is that regeneration of the liver after resection is a compensatory hyperplasia rather than a true restoration of the liver's original gross anatomy and architecture [21] (Figure 1A). The degree of hyperplasia is precisely controlled so that the process stops once an appropriate liver-to-bodyweight ratio has been achieved.

2.3. Hepatic Stem Cells and Second Line of Response to Liver Injury

Hepatic regeneration can be inhibited by several pathologic conditions. These include diabetes mellitus, malnutrition, aging, infection, chronic ethanol consumption, biliary obstruction and, more generally, chronic liver diseases. A common feature of all chronic liver diseases is progression to fibrosis, characterized by an increased production of matrix proteins, induced mainly by activated hepatic stellate cells, and a decreased matrix remodeling. With fibrosis there is usually diffuse inflammation and hepatocyte death, with evidence of an increase in the proportion of senescent hepatocytes, with the cell cycle arrested at G1/S transition. The rate of hepatocyte telomere shortening, which hampers cell division, has also been shown to correlate with the rate of progression of fibrosis [22]. Altogether, these data support the concept that the progression of liver fibrosis is associated with an impaired liver regeneration.

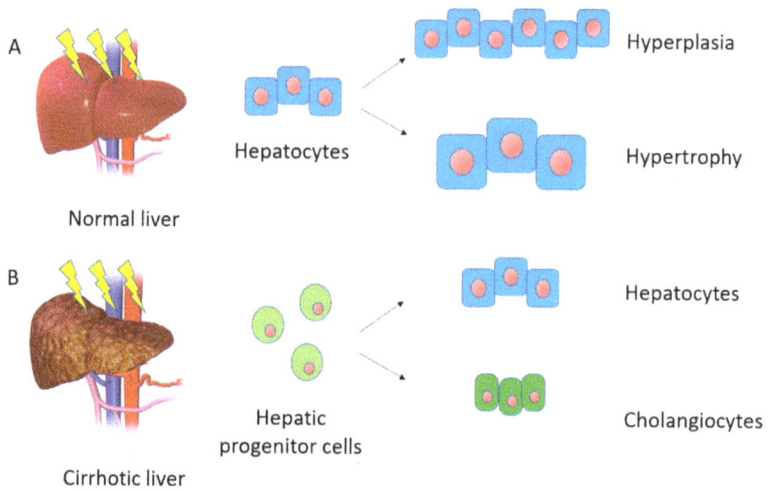

Figure 1. Schematic representation of mechanisms of liver regeneration. (**A**) After liver damage in normal conditions, the principal ways to restore hepatic mass are hyperplasia and hypertrophy. (**B**) In the cirrhotic liver, the normal regeneration process is impaired and hepatic progenitor cells are involved in restoring liver functions.

It has been hypothesized that, during chronic liver injury, when hepatocyte proliferation is impaired, the HPCs or oval cells orchestrate the regeneration process [23] (Figure 1B). The existence and regenerative potential of HPCs have been questioned, however. Farber was the first to report on the presence of a liver progenitor cell population in 1956, when he identified small cells with a high nucleus-to-cytoplasm ratio in the liver, that he called "oval cells" [24]. Subsequent works [25,26] demonstrated that these cells were activated in animal models of liver injury and had a bipotential ability to differentiate into hepatocytes and bile duct cells. Most of the data have come from animal models with the chemically-induced inhibition of native hepatocytes, in conjunction with the stimulation of liver regeneration. Lineage-tracing experiments have localized the adult human equivalent of these progenitor cells in the canal of Hering in the periportal regions of the hepatic lobules [27].

HPCs have the capacity to differentiate into hepatocytes and biliary cells in vitro, and to form hepatocyte buds, repopulating the damaged parenchyma in specific situations in vivo, in what is called "oval cell proliferation" in rodent models, and a "ductular reaction" in humans [28–30].

On activation, when the adult hepatocytes are unable to regenerate the injured liver, due either to senescence or cell cycle arrest, the HPCs proliferate in the portal zone and migrate towards the central vein in the liver lobules, gradually going through different states of maturity and function along the way, according to the so-called "streaming liver hypothesis" [29–31].

While the above is the most widely-accepted theory, work by Kuwahara et al. [32] suggests that it may be an oversimplification and that the liver might have a multi-tiered system of regeneration. There may be up to four potential stem cell niches in the canal of Hering, the intralobular bile ducts, the periductal mononuclear cells, and the peribiliary hepatocytes.

However, despite the accumulating evidence of HPC proliferation in liver injury, the extent of these cells' contribution to the natural history of human liver disease, and the triggers that activate this cell population are still not well understood.

2.4. Liver Regeneration, Inflammation, and Gender

Liver regeneration is closely linked to inflammation. Indeed, hepatic inflammation is a complex process originating in response to specific stress stimuli, which modulates the outcome of liver damage [33].

The inflammatory response can have both hepato-protective and detrimental effects. A controlled inflammatory reaction could be adjuvant to tissue regeneration, promoting the re-establishment of homeostasis. On the other hand, excessive and permanent inflammation could exacerbate the severity of hepatic parenchymal damage contributing to the irreversible decline of liver function [34]. Given its fundamental role, the inflammatory process is strictly controlled at a molecular and cellular level. Both resident (Kupffer) cells and circulating immune cells (lymphocytes and monocytes) are involved [33]. Among the molecular pathways involved in liver regeneration, IL-6 and IL-22 produced by activated natural killer (NK) and T cells in the liver induce activation of the signal transducer and activator of transcription 3 (STAT3) [35,36], while interferon-γ (IFN-γ) produced by B and T cells activates STAT1, inhibiting liver fibrosis and regeneration [37,38].

In a recent study [39], we demonstrated that a different immune response (in terms of the composition and maturation status of the cells involved) influence liver regeneration in males and females. The liver is known to be a gender-dimorphic organ in mammals, exhibiting sex-related differences in various aspects, such as the profile of steroid and drug metabolism [40], the number of hepatocytes and Kupffer cells [41], and the regeneration rate [42,43]. We demonstrated that female mice showed a more rapid recruitment of monocytes and $F4/80^{high}CD11b^{high}$ cells, and that the delay in recruitment of the same cells in male mice was controlled directly by the androgen receptor. Evidence from patients with drug-induced liver injury (DILI) confirms these observations, suggesting that males show a delay in regenerative response to an acute liver injury, possibly related to a maturation shift in monocytes. These findings might provide interesting starting points for new, gender-specific biomarkers, or for novel therapeutic interventions targeting monocyte recruitment or sex-hormone signaling [44,45]. Larger observational or prospective trials are needed, however, to better understand sex-dependent immune mechanisms in DILI.

3. Alternatives to Liver Transplantation

Liver transplantation (LT) is a widely-recognized treatment for patients with end-stage liver disease. Since the first success story reported by Thomas Starzl in 1967 [46], the short- and long-term outcomes of transplanted patients have gradually improved thanks to advances in the management of immunosuppressant therapies, more appropriate donor-recipient matching, and a better treatment of post-transplant comorbidities [47].

In recent years, we have witnessed an increasing number of patients worldwide on the waiting list for a transplant, but the number of available donors has not increased accordingly [48]. This gap between the patients needing a transplant and the donor organs available is a key issue in LT, with a mortality risk while on the waiting list of approximately 15% [49,50].

Liver regenerative medicine could cope with the donor shortage by using innovative approaches based on cell therapy and tissue/organ engineering. In the following sections, we briefly describe these ground-breaking alternatives to LT. Figure 2 summarizes the principal cell sources available for cell therapy and liver bioengineering, with their pros and cons.

3.1. Cell-Based Regeneration Therapy

As the demand for donor organs grows, therapeutic alternatives to liver transplantation must be sought. One such possible alternative is cell therapy, which may have two roles in the treatment of chronic liver diseases. Its first role is to control disease progression by stimulating endogenous regeneration and inhibiting fibrosis, thus ideally eliminating the need for liver transplantation [51]. When liver transplantation cannot be avoided, cell therapy may act as a bridge to surgery supporting liver function and, potentially, reducing the waitlist mortality rate. During the last ten years, hepatocytes, macrophages and stem cells have been transplanted with vary-ing degrees of success.

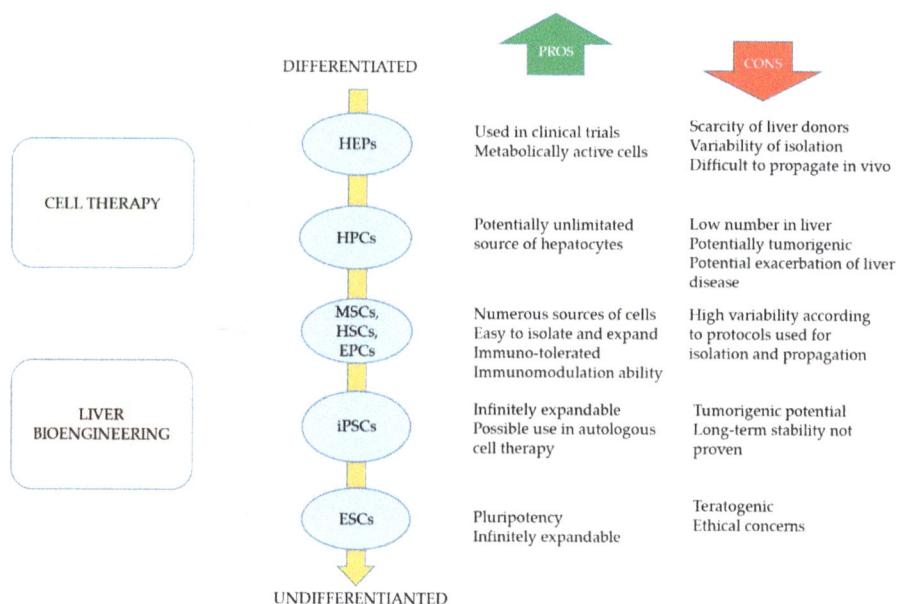

Figure 2. Schematic representation of the principal cell sources available for cell therapy and liver bioengineering, with brief description of their pros and cons. HEPs: hepatocytes; HPCs: hepatic progenitor cells; MSCs: mesenchymal stem cells; HSCs: hematopoietic stem cells; EPCs: endothelial progenitors cells; iPSCs: induced pluripotent stem cells; ESCs: embryonic stem cells.

3.1.1. Hepatocytes

Primary hepatocytes are the cells traditionally used for cell therapy in chronic liver diseases. It has been demonstrated that splenic or portal vein infusions of hepatocytes could induce modest reductions in ammonia levels and encephalopathy in both animal models and humans [52]. However, there are several important limitations to the use of human hepatocytes in the treatment of chronic liver diseases. One of the most important drawbacks is the difficulty of isolating a sufficient quantity of high-quality, metabolically-active cells. Hepatocytes are typically harvested from livers not suitable for transplantation, with a consequent variability in their quantity and quality [53]. Hepatocytes also rapidly lose their proliferative ability when cultured in vitro, and they are sensitive to freeze-thaw damage so their viability and engraftment are affected by culture and cryopreservation methods [54]. Innovative technologies that can expand, maintain, mature, and create hepatocytes in vitro, or alternative sources of cells are consequently required for future cell-based therapies for liver diseases.

3.1.2. Macrophages

Evidence has emerged from numerous human and animal studies of liver fibrosis being a two-way process and potentially reversible. The main regulator of this dynamic fibrogenesis-fibrosis resolution paradigm seems to be the hepatic macrophage [55,56]. This apparently dichotomous effect of macrophages in liver fibrosis is attributable to the balance of profibrotic and restorative macrophages [57]. A better understanding of the mechanisms controlling this process could yield novel monocyte/macrophage-based cell therapies.

Technological advances in the stem-cell field could lead to therapeutic approaches based on the autologous propagation of monocytic populations, or possibly their derivation from embryonic stem cell components. A monocyte/macrophage-based approach to damping liver fibrosis has already been attempted in animal models. Thomas et al. [58] examined the therapeutic potential of exogenous bone marrow (BM) cells, and those of the monocyte-macrophage lineage in particular, in a mice model of chronic liver injury. They found that the intraportal administration of differentiated BM-derived macrophages (BMMs) improved liver fibrosis, regeneration, and function via a wide range of reparative pathways, with a therapeutic benefit. On the other hand, liver fibrosis was not significantly ameliorated by the infusion of macrophage precursors, and it was even exacerbated by whole BM cells. Thanks to paracrine signaling from the BMMs to larger populations of endogenous cells, their effect was amplified. As a consequence, a modest number of donor BMMs could exert whole-organ changes—encouraging a translational perspective and suggesting a future clinical potential.

3.1.3. Pluripotent Stem Cells

Embryonic stem cells (ESCs) are pluripotent cells derived from the inner cell mass of the blastocyst. These cell were first characterized in 1998 by Thomson et al. [59]. They have pluripotency and can potentially differentiate into all somatic cells [60]. Numerous studies have demonstrated the differentiation of ESCs into hepatocyte-like cells that express a number of hepatocyte-related genes and mimic liver function [61–65]. ESC-derived hepatocytes also have the typical morphology of mature hepatocytes and are able to colonize liver tissue after transplantation, promoting the injured liver's recovery via cell replacement and stimulating endogenous regeneration [63,66–68]. Despite these promising results and the favorable characteristics of human ESCs, such as a good resistance to cryopreservation, practical and ethical barriers have always precluded their application in clinical practice.

Human induced pluripotent stem cells (iPSCs) have recently emerged as a way of bypassing the ethical concerns associated with the use of ESCs [69]. The iPSCs are derived by reprogramming mature somatic cells induced by different transcription factors [70]. Their characteristics of self-renewal and pluripotency make iPSCs good substitutes for ESCs, and an appealing source of normal human cells that can differentiate into virtually any somatic cell type, including hepatocytes. The hepatocyte-like cells (HLCs) derived from human iPSCs could provide a stable source of hepatocytes for multiple applications, including cell therapy, disease modelling, and drug safety screening [71,72]. Protocols adopted to differentiate human ESCs and human iPSCs into HLCs mimic the developmental pathway of the liver during embryogenesis, and have vastly improved in recent years. Nevertheless, several issues regarding the safety and reproducibility of iPSCs still need to be settled before their real clinical application, including tumorigenicity and teratoma formation, the debate on their immunogenicity, long-term safety and efficacy, and the optimal reprogramming and manufacturing processes [73–75]. Constant progress is nonetheless being made in reprogramming technologies, and in new and improved manufacturing methods.

3.1.4. Adult Stem Cells

Stem cells are valid alternative sources of cells for the treatment of liver diseases. They could potentially be involved in modulating the liver's regenerative processes to reduce scarring in cirrhosis, and to down-regulate immune-mediated liver damage. Stem cells could also be differentiated into hepatocytes for cell transplantation, or used in extracorporeal bioartificial liver systems [76].

Different types of adult stem cells have been tested over the years, including hematopoietic stem cells (HSCs), mesenchymal stem cells (MSCs), endothelial progenitor cells (EPCs), and hepatic progenitor cells (HPCs) [77–80].

HSCs are the predominant population of stem cells in bone marrow, and express the surface marker CD34. HSCs can easily be isolated in the bloodstream after treatment with mobilizing agents, the most widely-studied and often-used of which is the granulocyte-colony stimulating factor [81]. Hepatocyte-like cells derived from HSCs have been demonstrated to support liver regeneration [82,83]. Different mechanisms have been suggested, such as the de novo generation of hepatocytes through transdifferentiation or the genetic reprogramming of resident hepatocytes through cell fusion [84,85]. However, the most plausible hypothesis is that the clinical benefit of HSC therapy occurs through paracrine signaling interactions involving various cytokines and growth factors, that stimulate regeneration and neoangiogenesis [86,87].

Endothelial progenitor cells (EPCs) can be found in both peripheral blood vessels and bone marrow, and their main function is to participate in the neovascularization of damaged tissue [88,89]. In the context of cell therapy for liver diseases, one animal study demonstrated that the transplantation of EPCs led to a lessening of liver fibrosis [79]. ESCs are also able to promote hepatocyte proliferation and increase matrix metalloproteinase activity [90]. All these effects are related to an increased secretion of specific growth factors [91,92].

Another promising cell treatment for liver diseases is based on mesenchymal stem cells (MSCs), a population of multipotent progenitors capable of differentiating towards adipogenic, osteogenic and hepatogenic lineages, with a low immunogenicity [93]. Bone marrow is considered the main source of MSCs [94], but alternative sources are being examined, such as adipose tissue [95], placenta, amniotic fluid, umbilical cord blood, and umbilical cord [96,97]. Our research has focused on umbilical cord MSCs. We demonstrated in an animal model that, when systematically administered, these cells can repair acute liver injury [97]. The ability of the same cells to repair tissue damage was also demonstrated in a chemically-induced intestinal injury in immunodeficient mice [98]. We and other authors have demonstrated that MSCs have the capacity to provide both metabolic and trophic support due to their potential for hepatocytic differentiation, and their secretion of anti-inflammatory, anti-apoptotic, immunomodulatory, and pro-proliferative factors [97–99]. This leads to liver function being restored via the repair of damaged tissue, the suppression of inflammation, and the stimulation of endogenous regeneration through paracrine effects [100].

Cell-based therapy using HPCs could potentially regenerate the liver during chronic diseases. Multiple protocols have been established for isolating HPCs in fetal and rodent models, and cell differentiation protocols are available for progenitor cells derived from the human liver or biliary tree [101–103]. Due to the low number of these cells in the liver, the use of autologous HPCs is probably unfeasible. The use of expanded fetal or syngeneic HPCs is more likely, though this approach raises questions regarding the engraftment rate of transplanted cells, and the need for immunosuppressant therapy. Despite the theoretical feasibility of such approaches, we still have only a limited understanding of HPCs, their precise role in liver pathophysiology, and how the entire process of regeneration/differentiation is regulated. Given the possible disadvantages of HPC activation, which might exacerbate disease progression or prompt the onset of cancer [104], all these issues warrant further study and careful examination before any therapeutic approaches could be applicable.

3.1.5. Hepatic Organoids

Considered as a bridge between liver cell therapy and liver bioengineering, hepatic organoids are functional three-dimensional (3D) in vitro models of the liver consisting of a spherical monolayer of epithelium that preserves the key physiological features of the liver [105]. Liver organoids are typically obtained by isolating and expanding stem cells or hepatic progenitor cells.

Liver organoids show a limited spontaneous differentiation during maintenance and expansion. For this reason, protocols for establishing organoids were divided into two steps. The first relied on proliferation culture conditions for the establishment and expansion of hepatic organoids. Then, in a second step, proliferative signals were removed, and differentiation towards hepatocyte-like cells was induced. These culture conditions enabled organoids to be obtained with 30–50% fulfilling hepatic characteristics [106], but without the complete functional repertoire of adult hepatocytes—a drawback shared by HPC-to-hepatocyte differentiation.

Differentiated hepatic organoids transplanted into mouse models of liver failure have demonstrated a capacity for engraftment and repopulation of the damaged liver, with partial rescue of liver function [105]. Equivalent human liver organoids transplanted into mice with acute liver damage were able to produce human albumin and alpha-1-antitrypin, with secretion levels comparable with those after the transplantation of adult hepatocytes [101].

Three-dimensional liver tissue has also been engineered by using human iPSCs to derive hepatocytes in co-culture with mesenchymal and endothelial cells [107]. When transplanted into mice, these liver buds were vascularized and matured to synthesize serum proteins and carry out detoxifying functions.

Current research is aiming for the clinical application of liver buds suitable for hepatic administration via the portal vein in patients in need of a liver transplant [108]. Among the different cell sources, adult stem cells directly derived from hepatic tissue are preferred. Indeed, drawbacks of human iPSCs or trans-differentiated cells used in the design of clinical solutions concern their exposure to genetic modifications through reprogramming factors, and their genomic instability, particularly in long-term cultures [109].

Moreover, liver organoids provide a novel platform for research on: 1) liver development and regeneration; 2) detoxification and metabolism; 3) liver disease modelling; and 4) adult stem cell biology.

3.2. Liver Tissue Bioengineering

Tissue engineering could offer various solutions for reducing the waiting list by creating biocompatible scaffolds and extracorporeal liver devices suitable for either in vitro or in vivo applications [110].

In the last two decades, a growing number of studies demonstrated that 3D cultures have a number of advantages over traditional two-dimensional (2D) cell cultures [110,111]. A physiologically 3D microenvironment is crucial to the development of in vitro tissue models, particularly for such complex tissues as the liver, in which the interaction between hepatocytes, hepatic stellate cells, and extracellular matrix (ECM) creates the microenvironment of the hepatic lobules [112].

The search for efficient biocompatible scaffolds aims to create organic or polymeric constructs that mimic the liver ECM and replicate functional characteristics such as cell adhesion, viability, growth, and proliferation. The principal strategies are based on biomaterials such as polymer-based 3D constructs, decellularized ECM, or bioprinting 3D constructs.

Another recent approach involves the development of bioreactors to improve various functions of hepatocytes that are seeded in constructs. In bioreactors, a real 3D microenvironment niche is created to improve cell attachment, growth, and proliferation, with a marked improvement in liver metabolism and function [110]. A more sophisticated technology is the liver-on-chip: A combination of bio-reactor techniques and microfluidic devices to sustain the phenotype of hepatocytes and liver-specific functions in long-term culture [113].

Below we provide an overview of such bioengineering approaches, and Figure 3 shows the main pros and cons of each of them.

3.2.1. Decellularized Extracellular Matrix

A new approach to liver regenerative medicine involves generating 3D organs with a decellularized, native liver bioscaffold that can be repopulated with parenchymal and non-parenchymal cells [114]. The liver's native ECM has a complex composition and topography, serving as a structure for cell-ECM adhesion, interaction, and polarity, with implications for the regulation of cell morphology, proliferation, differentiation, and viability interactions [115]. Donor organs unsuitable for transplantation are used to create whole-liver scaffolds which are subsequently reseeded with healthy cells to create transplantable liver grafts. The scaffolds maintain the native liver architecture and ECM composition, which allows for proper cell homing and function. Decellularization techniques were introduced in the 1980s [116], but the concept of whole-organ decellularization was developed later by Ott and colleagues in mice hearts [117]. This technique was later adapted for liver engineering purposes [118], with the preservation of the chemical composition and structure of the ECM with structurally intact vessels, and bile ducts. This bioscaffold was then recellularized with hepatocytes and endothelial cells. The recellularized graft transplanted in vivo and perfused ex vivo demonstrated mature liver functions. Further improvements in the technique were obtained over the years, such us multistep cell seeding, the use of stem cells (MSCs, fetal hepatocytes, iPSCs) [119–121], optimization of the decellularization cocktail, and perfusion without any thrombus formation [122]. The feasibility of this technique was also demonstrated in larger animal models [123], and even in humans [124], bringing the approach to clinical scale.

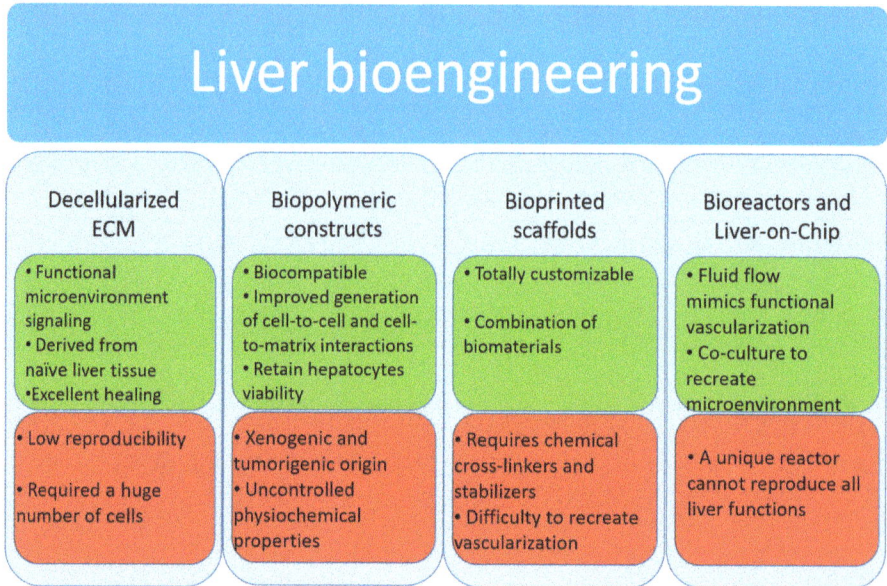

Figure 3. Main pros (green boxes) and cons (red boxes) of the principal liver bioengineering approaches. ECM: extracellular matrix.

All these studies demonstrated that decellularized livers hold great potential as a therapeutic approach, but numerous pitfalls remain. First, the technique allows for the successful seeding and culture of hepatocytes, but colonization of the bile duct with functional cells and the achievement of an intact vascular network remain to be perfected. Another important issue before whole liver bioscaffolds can be used in clinical practice is the lack of a suitable source of cells, which should be readily available and renewable because successful liver recellularization demands hundreds of millions of cells. The limited availability and inability to expand primary hepatocytes has led researchers in the field to search for a new cell source. Although many groups have attempted to overcome the problem by using fetal liver cells, stem cells or iPSCs, the production of such huge numbers of hepatocytes is still far beyond current technical capability.

Another hurdle that should be promptly addressed is "sample to sample" variation due to the unique condition of each donor deriving from the use of discarded livers [125]. The next goals of bioengineering research will be to solve these problems.

3.2.2. Biopolymer Constructs

In modern tissue engineering, efforts are being made to make natural biomaterials mimic the natural hepatic ECM. The main components of these scaffolds are collagen and hyaluronic acid. The latter strongly supports cell attachment, proliferation, differentiation, growth, and migration. Immature and mature hepatocytes express CD44, the surface receptors for hyaluronic acid, so biopolymers with hyaluronic acid and its derivatives have more adhesive power for hepatocytes. They can retain hepatocyte viability for 4 weeks [126].

Other natural biomaterials used in the construction of bioactive scaffolds are alginate, chitin, chitosan, silk, Matrigel®, and sponge. Matrigel® is a scaffold consisting of a mixture of ECM proteins derived from the basal membranes of murine chondrosarcoma, which contains laminin, heparan sulfate proteoglycan, and collagen type IV [127]. It has been used in numerous studies to culture hepatocytes and induce the hepatic differentiation of stem cells [97,128].

Although hydrogels formed by natural biomaterials such as alginate and Matrigel® are biocompatible and improve the generation of cell-to-cell and cell-to-matrix interactions, they have some important limits that prevent their clinical application. The main shortcomings of such biomaterials are their uncontrollable physicochemical properties, degradability, lack of regenerative ability, and inconsistent mechanical properties. Moreover, due to the xenogenic and tumorigenic origin of Matrigels, they are not an optimal support for clinical applications in liver bioengineering [129].

By comparison with natural biomaterials, synthetic materials offer a wide range of properties and a better control over them. Scaffolds containing biodegradable polymers, such as polylactic acid, polyglycolic acid, polyanhydrides, polyfumarates, polyorthoesters, polycaprolactones, poly-L-lactic acid, and polycarbonates facilitate cell regeneration, transplantation, and degradation on time [130]. The biocompatibility of bioengineered matrices and scaffold adhesion properties could also be improved by chemically modifying these polymers (e.g., by incorporating proteins and special bioactive domains), stimulating cell attachment and migration, and thereby facilitating liver tissue repair [131].

While natural and synthetic materials support the successful culture of hepatocytes, these constructs fail to perfectly reproduce the microenvironment of the liver essential to a functional liver cell activity. For this reason, their therapeutic potential is limited.

3.2.3. Bioprinted Scaffolds

Although the use of biomaterials in 3D culture has improved the settings for liver tissue engineering, it has some limitations. These include the difficulty of creating complex biological structures and designs due to size, material, compositional, and technological constraints [132]. An innovative solution to these problems involves using bioprinted scaffolds, tissue-mimicking constructs created by means of a bioprinting process with biocompatible materials (i.e., bio-inks) [133]. Advances in bioprinting technology have enabled the creation of more complex 3D structures using combinations of different biomaterials and cell types [134]. The chance to totally customize the prints also guarantees the complete personalization of such scaffolds and their applications. The available bioprinting modalities include extrusion, inkjet, and laser-assisted bioprinting [135]. Extrusion bioprinting, the most often-used bioprinting modality in biomedical research, allows for a strong degree of customization with few restrictions on the cells used [134]. The choice of biomaterials is more restrictive, however, as they are either easy to print or ideal for cell culture, but typically not both [136]. The ideal characteristics of bio-inks for extrusion bioprinting are viscosity to enable printing, associated with an adequate elasticity to maintain their structure, while also maintaining cell viability and supporting cell function [132].

The most common biomaterials used for bioprinting are collagen, alginate, polyethylene glycol (PEG), hyaluronic acid, fibrin, gelatin, or polycaprolactone, each with unique properties [133]. With the exception of collagen, these biomaterials need the addition of a cross-linker that could adversely affect the cells. For this reason, they should be appropriately balanced to guarantee the best biocompatibility of the bio-ink being used [133]. Although collagen is an ideal material for in-vivo-like tissue replication, it is a poor bio-ink because it has a time- and temperature-sensitive cross-linking [137]. A multi-component hybrid bio-ink is therefore a potential solution for achieving ideal physiological relevance and bio-printability. Unfortunately, durable 3D construct fabrication requires the incorporation of chemical stabilizers, such as polycaprolactone, showing the limitations of bio-inking technologies in mimicking both the biochemical composition and the complex 3D structure of the liver.

Another important challenge in 3D bioprinting is how to fabricate and mimic cellular microenvironments from molecular to macroscopic scales for tissue engineering and regenerative medicine. Using this approach, the researcher aims to create a whole functional liver suitable for transplantation, but some important issues, such as vascularization, should be addressed before this methodology can really be implemented.

3.3. Bioreactor Systems

Despite the great progress made in biomaterial development for tissue engineering, some challenges need to be overcome. The most important limiting parameter in tissue engineering and bioprinting concerns vascularization [138]. Without suitable vascularization, cells are subject to hypoxia, toxemia, apoptosis, and immediate cell death. The bioreactor approach aims to overcome this limitation. In fact, the bioreactor involves a designed or programmed fluid flow as an integral part of the culture format. The flow in perfusion bioreactors enables a continuous exchange of nutrients, a better oxygen delivery, and a physiological shear stress, influencing cell function in ways that are impossible to achieve in static culture formats [139].

The evolution of bioreactor technologies has paralleled advances in the development of functional biomaterial scaffolds [140]. The scaffold not only provides an adhesion surface for cells, but also profoundly influences cell shape and gene expression relevant to cell growth and liver-specific functions. Moreover, when placed as a separation between cells and the medium, the scaffolds act as a modulator for water and nutrient transport from the medium to the cells, and discharge waste metabolites from the cells to the medium [141].

Four principal types of bioreactors have been used for liver cell culture: 1) flat plate and monolayer; 2) hollow fiber; 3) perfused beds and scaffolds; and 4) encapsulation and suspension. With the exception of type 1, the other bioreactors enable the 3D monoculture or co-culture of hepatocytes under tissue-specific mechanical forces (pressure, shear stress, flow) [142,143]. Some of these bioreactors have been used as bioartificial livers, charged with various types of liver cells, as a bridge for patients with acute liver failure awaiting transplantation [144]. Now the challenge is to use cell-based bioreactors as in vitro screening systems for drug toxicity, metabolism evaluation and potential clinical treatments.

Some parameters are crucial to hepatocyte vitality and functionality, including various biophysical factors such as oxygenation, hemodynamics, and shear stress. Perfusion in bioreactor devices enables the establishment of oxygen gradients and hepatic zonation, resulting in graded CYP expression and metabolism [145,146]. A controlled oxygen gradient from 25 to 70 mmHg inside a hepatic bioreactor creates a functional hepatocyte zonation similar to what is observed in vivo. Cell oxygenation could be partially controlled by varying the medium flow rate, but may consequently exert a shear stress on the hepatocytes. Flow rate should be carefully controlled since cell damage can occur. Hydrodynamic stress induces ECM remodeling, scaffold degradation and changes in tissue composition, influencing the device's structural and mechanical properties. On the other hand, low flow rates limit the oxygen supply, lead to nutrient deficiency, and reduce cell viability and survival probability [146].

The co-culture of hepatocytes and non-parenchymal cells is important for the reorganization of hepatocytes in culture by secreting cytokines, nitric oxide, and matrix components [145,147]. Co-culture is also useful for inducing liver-specific functions, preserving maximal levels of functional adhesion molecule expression, and reducing the number of cells needed for a bioartificial liver [144].

The main limitation of the bioreactors is that not all critical liver functions can be replicated on the desired level as yet. For this reason, based on the present state of the art, a unique bioreactor that can faithfully reproduce all liver functions is still lacking.

3.4. Micro-Bioreactors and Liver-on-Chip

The combination of nanotechnology, microchips, and microfluidics in a single device has great potential for applications in liver tissue engineering. Various strategies have been developed to obtain micro-bioreactors. Microsystems technology has been used to fabricate 2D or 3D culture devices by using different types of materials, like silicon, silicone elastomer, and biocompatible and biodegradable polymers. Such systems typically exhibit laminar flow, similar to the environment in vivo, and allow the creation of microfluidic channels with larger surface-to-volume ratios suitable for oxygen and nutrition supply [113].

Other interesting systems that exploit microfluidic technology are the so-called "liver-on-chip" devices [148]. These systems consist of microchambers containing engineered tissue and living cell cultures interconnected by a microfluidic network. Such organs on chips enable the study of human phys-iology in an organ-specific context, and the development of novel in vitro disease models. They have the potential to serve as replacements for animals used in drug development, toxin testing, and screening for biothreats and chemical warfare agents [149].

4. Conclusions

In conclusion, regenerative medicine and bioengineering are cutting-edge technologies that look promising as a final solution to the treatment of end-stage liver diseases. A better understanding of liver regeneration and the development of in vitro systems that successfully mimic hepatocyte expansion and differentiation will make autologous cell therapy a feasible alternative to liver transplantation. The current scenario is also moving towards the successful development of whole bioengineered livers and their effective use in clinical practice in lieu of liver transplantation.

Author Contributions: Writing—original draft preparation: D.B.; writing—review and editing: F.R.P. and P.B.

Funding: This research received no external funding.

Acknowledgments: Authors are grateful to Marina Minnaja Foundation for co-funding Debora Bizzaro post-doctoral fellowship.

Conflicts of Interest: The authors declare no conflict of interest.

References

1. Manns, M.P.; Burra, P.; Sargent, J.; Horton, R.; Karlsen, T.H. The Lancet-EASL Commission on liver diseases in Europe: overcoming unmet needs, stigma, and inequities. *Lancet* **2018**, *392*, 621–622. [CrossRef]
2. Germani, G.; Theocharidou, E.; Adam, R.; Karam, V.; Wendon, J.; O'Grady, J.; Burra, P.; Senzolo, M.; Mirza, D.; Castaing, D.; et al. Liver transplantation for acute liver failure in Europe: Outcomes over 20 years from the ELTR database. *J. Hepatol.* **2012**, *57*, 288–296. [CrossRef] [PubMed]
3. Burra, P.; Berenguer, M.; Pomfret, E. The ILTS Consensus Conference on NAFLD/NASH and liver transplantation. *Transplantation* **2018**. [CrossRef]
4. Pang, Y.; Horimoto, Y.; Sutoko, S.; Montagne, K.; Shinohara, M.; Mathiue, D.; Komori, K.; Anzai, M.; Niino, T.; Sakai, Y. Novel integrative methodology for engineering large liver tissue equivalents based on three-dimensional scaffold fabrication and cellular aggregate assembly. *Biofabrication* **2016**, *8*, 16. [CrossRef] [PubMed]
5. Su, Z.C.; Li, P.S.; Wu, B.G.; Ma, H.; Wang, Y.C.; Liu, G.X.; Zeng, H.L.; Li, Z.Z.; Wei, X. PHBVHHx scaffolds loaded with umbilical cord-derived mesenchymal stem cells or hepatocyte-like cells differentiated from these cells for liver tissue engineering. *Mater. Sci. Eng. C Mater. Biol. Appl.* **2014**, *45*, 374–382. [CrossRef]
6. Tremblay, K.D.; Zaret, K.S. Distinct populations of endoderm cells converge to generate the embryonic liver bud and ventral foregut tissues. *Dev. Biol.* **2005**, *280*, 87–99. [CrossRef]
7. Houssaint, E. Differentiation of the mouse hepatic primordium. I. An analysis of tissue interactions in hepatocyte differentiation. *Cell Differ.* **1980**, *9*, 269–279. [CrossRef]
8. Medlock, E.S.; Haar, J.L. The liver hemopoietic environment: I. Developing hepatocytes and their role in fetal hemopoiesis. *Anat. Rec.* **1983**, *207*, 31–41. [CrossRef] [PubMed]
9. Lemaigre, F.P. Development of the biliary tract. *Mech. Dev.* **2003**, *120*, 81–87. [CrossRef]
10. Kung, J.W.C.; Currie, I.S.; Forbes, S.J.; Ross, J.A. Liver Development, Regeneration, and Carcinogenesis. *J. Biomed. Biotechnol.* **2010**. [CrossRef]
11. Tanimizu, N.; Miyajima, A. Notch signaling controls hepatoblast differentiation by altering the expression of liver-enriched transcription factors. *J. Cell Sci.* **2004**, *117*, 3165–3174. [CrossRef] [PubMed]
12. McCright, B.; Lozier, J.; Gridley, T. A mouse model of Alagille syndrome: Notch2 as a genetic modifier of Jag1 haploinsufficiency. *Development* **2002**, *129*, 1075–1082. [PubMed]

13. Suzuki, A.; Iwama, A.; Miyashita, H.; Nakauchi, H.; Taniguchi, H. Role for growth factors and extracellular matrix in controlling differentiation of prospectively isolated hepatic stem cells. *Development* **2003**, *130*, 2513–2524. [CrossRef] [PubMed]

14. Jochheim, A.; Cieslak, A.; Hillemann, T.; Cantz, T.; Scharf, J.; Manns, M.P.; Ott, M. Multi-stage analysis of differential gene expression in BALB/C mouse liver development by high-density microarrays. *Differentiation* **2003**, *71*, 62–72. [CrossRef] [PubMed]

15. Zhao, R.; Duncan, S.A. Embryonic development of the liver. *Hepatology* **2005**, *41*, 956–967. [CrossRef] [PubMed]

16. Forbes, S.J.; Rosenthal, N. Preparing the ground for tissue regeneration: from mechanism to therapy. *Nat. Med.* **2014**, *20*, 857–869. [CrossRef]

17. Riehle, K.J.; Dan, Y.Y.; Campbell, J.S.; Fausto, N. New concepts in liver regeneration. *J. Gastroenterol. Hepatol.* **2011**, *26*, 203–212. [CrossRef]

18. MacDonald, R.A. Lifespan of liver cells: autoradiographic study using tritiated thymidine in normal, cirrhotic, and partially hepatectomized rats. *Arch. Int. Med.* **1961**, *107*, 335–343. [CrossRef]

19. Gilgenkrantz, H.; Collin de l'Hortet, A. New insights into liver regeneration. *Clin. Res. Hepatol. Gastroenterol.* **2011**, *35*, 623–629. [CrossRef]

20. Michalopoulos, G.K. Hepatostat: Liver Regeneration and Normal Liver Tissue Maintenance. *Hepatology* **2017**, *65*, 1384–1392. [CrossRef]

21. Taub, R. Liver regeneration: From myth to mechanism. *Nat. Rev. Mol. Cell Biol.* **2004**, *5*, 836–847. [CrossRef] [PubMed]

22. Aravinthan, A.; Scarpini, C.; Tachtatzis, P.; Verma, S.; Penrhyn-Lowe, S.; Harvey, R.; Davies, S.E.; Allison, M.; Coleman, N.; Alexander, G. Hepatocyte senescence predicts progression in non-alcohol-related fatty liver disease. *J. Hepatol.* **2013**, *58*, 549–556. [CrossRef] [PubMed]

23. Fausto, N.; Campbell, J.S.; Riehle, K.J. Liver regeneration. *J. Hepatol.* **2012**, *57*, 692–694. [CrossRef]

24. Farber, E. Similarities in the sequence of early histological changes induced in the liver of the rat by ethionine, 2-acetylamino-fluorene, and 3'-methyl-4-dimethylaminoazobenzene. *Cancer Res.* **1956**, *16*, 142–148. [PubMed]

25. Lazaro, C.A.; Rhim, J.A.; Yamada, Y.; Fausto, N. Generation of hepatocytes from oval cell precursors in culture. *Cancer Res.* **1998**, *58*, 5514–5522. [PubMed]

26. Dunsford, H.A.; Karnasuta, C.; Hunt, J.M.; Sell, S. Different lineages of chemically-induced hepatocellular-carcinoma in rats defined by monoclonal-antibodies. *Cancer Res.* **1989**, *49*, 4894–4900. [PubMed]

27. Theise, N.D.; Saxena, R.; Portmann, B.C.; Thung, S.N.; Yee, H.; Chiriboga, L.; Kumar, A.; Crawford, J.M. The canals of Hering and hepatic stem cells in humans. *Hepatology* **1999**, *30*, 1425–1433. [CrossRef]

28. Stueck, A.E.; Wanless, I.R. Hepatocyte Buds Derived From Progenitor Cells Repopulate Regions of Parenchymal Extinction in Human Cirrhosis. *Hepatology* **2015**, *61*, 1696–1707. [CrossRef]

29. Roskams, T.A.; Theise, N.D.; Balabaud, C.; Bhagat, G.; Bhathal, P.S.; Bioulac-Sage, P.; Brunt, E.M.; Crawford, J.M.; Crosby, H.A.; Desmet, V.; et al. Nomenclature of the finer branches of the biliary tree: Canals, ductules, and ductular reactions in human livers. *Hepatology* **2004**, *39*, 1739–1745. [CrossRef]

30. Zajicek, G.; Oren, R.; Weinreb, M. THE STREAMING LIVER. *Liver* **1985**, *5*, 293–300. [CrossRef]

31. Roskams, T.; Yang, S.Q.; Koteish, A.; Durnez, A.; DeVos, R.; Huang, X.W.; Achten, R.; Verslype, C.; Diehl, A.M. Oxidative stress and oval cell accumulation in mice and humans with alcoholic and nonalcoholic fatty liver disease. *Am. J. Pathol.* **2003**, *163*, 1301–1311. [CrossRef]

32. Kuwahara, R.; Kofman, A.V.; Landis, C.S.; Swenson, E.S.; Barendswaard, E.; Theise, N.D. The hepatic stem cell niche: Identification by label-retaining cell assay. *Hepatology* **2008**, *47*, 1994–2002. [CrossRef] [PubMed]

33. Kubes, P.; Mehal, W.Z. Sterile Inflammation in the Liver. *Gastroenterology* **2012**, *143*, 1158–1172. [CrossRef] [PubMed]

34. Schattenberg, J.M.; Galle, P.R.; Schuchmann, M. Apoptosis in liver disease. *Liver Int.* **2006**, *26*, 904–911. [CrossRef] [PubMed]

35. Haga, S.; Terui, K.; Zhang, H.Q.; Enosawa, S.; Ogawa, W.; Inoue, H.; Okuyama, T.; Takeda, K.; Akira, S.; Ogino, T.; et al. Stat3 protects against Fas-induced liver injury by redox-dependent and -independent mechanisms. *J. Clin. Investig.* **2003**, *112*, 989–998. [CrossRef]

36. Li, W.; Liang, X.P.; Kellendonk, C.; Poli, V.; Taub, R. STAT3 contributes to the mitogenic response of hepatocytes during liver regeneration. *J. Biol. Chem.* **2002**, *277*, 28411–28417. [CrossRef] [PubMed]

37. Jeong, W.I.; Park, O.; Radaeva, S.; Gao, B. STAT1 inhibits liver fibrosis in mice by inhibiting stellate cell proliferation and stimulating NK cell cytotoxicity. *Hepatology* **2006**, *44*, 1441–1451. [CrossRef]

38. Sun, Z.L.; Klein, A.S.; Radaeva, S.; Hong, F.; El-Assal, O.; Pan, H.N.; Jaruga, B.; Batkai, S.; Hoshino, S.; Tian, Z.G.; et al. In vitro interleukin-6 treatment prevents mortality associated with fatty liver transplants in rats. *Gastroenterology* **2003**, *125*, 202–215. [CrossRef]

39. Bizzaro, D.; Crescenzi, M.; Di Liddo, R.; Arcidiacono, D.; Cappon, A.; Bertalot, T.; Amodio, V.; Tasso, A.; Stefani, A.; Bertazzo, V.; et al. Sex-dependent differences in inflammatory responses during liver regeneration in a murine model of acute liver injury. *Clin. Sci. (Lond.)* **2018**, *132*, 255–272. [CrossRef]

40. Yates, F.E.; Herbst, A.L.; Urquhart, J. Sex difference in rate of ring A reduction of delta 4–3-keto-steroids in vitro by rat liver. *Endocrinology* **1958**, *63*, 887–902. [CrossRef]

41. Marcos, R.; Lopes, C.; Malhao, F.; Correia-Gomes, C.; Fonseca, S.; Lima, M.; Gebhardt, R.; Rocha, E. Stereological assessment of sexual dimorphism in the rat liver reveals differences in hepatocytes and Kupffer cells but not hepatic stellate cells. *J. Anat.* **2016**, *228*, 996–1005. [CrossRef] [PubMed]

42. Tsukamoto, I.; Kojo, S. The sex difference in the regulation of liver regeneration after partial hepatectomy in the rat. *Biochim. Biophys. Acta* **1990**, *1033*, 287–290. [CrossRef]

43. Imamura, H.; Shimada, R.; Kubota, M.; Matsuyama, Y.; Nakayama, A.; Miyagawa, S.; Makuuchi, M.; Kawasaki, S. Preoperative portal vein embolization: an audit of 84 patients. *Hepatology* **1999**, *29*, 1099–1105. [CrossRef] [PubMed]

44. Francavilla, A.; Eagon, P.K.; DiLeo, A.; Polimeno, L.; Panella, C.; Aquilino, A.M.; Ingrosso, M.; Van Thiel, D.H.; Starzl, T.E. Sex hormone-related functions in regenerating male rat liver. *Gastroenterology* **1986**, *91*, 1263–1270. [CrossRef]

45. Yamaguchi, M.; Yu, L.; Nazmy El-Assal, O.; Satoh, T.; Kumar Dhar, D.; Yamanoi, A.; Nagasue, N. Androgen metabolism in regenerating liver of male rats: evidence for active uptake and utilization of testosterone. *Hepatol. Res.* **2001**, *20*, 114–127. [CrossRef]

46. Starzl, T.E.; Marchioro, T.L.; Porter, K.A.; Brettschneider, L. Homotransplantation of the liver. *Transplantation* **1967**, *5*, 790–803. [CrossRef] [PubMed]

47. Adam, R.; Karam, V.; Delvart, V.; O'Grady, J.; Mirza, D.; Klempnauer, J.; Castaing, D.; Neuhaus, P.; Jamieson, N.; Salizzoni, M.; et al. Evolution of indications and results of liver transplantation in Europe. A report from the European Liver Transplant Registry (ELTR). *J. Hepatol.* **2012**, *57*, 675–688. [CrossRef] [PubMed]

48. Dutkowski, P.; Linecker, M.; DeOliveira, M.L.; Mullhaupt, B.; Clavien, P.A. Challenges to Liver Transplantation and Strategies to Improve Outcomes. *Gastroenterology* **2015**, *148*, 307–323. [CrossRef]

49. Kim, W.R.; Therneau, T.M.; Benson, J.T.; Kremers, W.K.; Rosen, C.B.; Gores, G.J.; Dickson, E.R. Deaths on the liver transplant waiting list: An analysis of competing risks. *Hepatology* **2006**, *43*, 345–351. [CrossRef]

50. Toniutto, P.; Zanetto, A.; Ferrarese, A.; Burra, P. Current challenges and future directions for liver transplantation. *Liver Int.* **2017**, *37*, 317–327. [CrossRef] [PubMed]

51. Bartlett, D.C.; Newsome, P.N. Hepatocyte cell therapy in liver disease. *Expert Rev. Gastroenterol. Hepatol.* **2015**, *9*, 1261–1272. [CrossRef]

52. Fisher, R.A.; Strom, S.C. Human hepatocyte transplantation: Worldwide results. *Transplantation* **2006**, *82*, 441–449. [CrossRef] [PubMed]

53. Huebert, R.C.; Rakela, J. Cellular Therapy for Liver Disease. *Mayo Clin. Proc.* **2014**, *89*, 414–424. [CrossRef] [PubMed]

54. Ferrer, J.R.; Chokechanachaisakul, A.; Wertheim, J.A. New Tools in Experimental Cellular Therapy for the Treatment of Liver Diseases. *Curr. Transplant. Rep.* **2015**, *2*, 202–210. [CrossRef] [PubMed]

55. Duffield, J.S.; Forbes, S.J.; Constandinou, C.M. Selective depletion of macrophages reveals distinct, opposing roles during liver injury and repair. *J. Clin. Investig.* **2005**, *115*, 56–65. [CrossRef] [PubMed]

56. Fallowfield, J.A.; Mizuno, M.; Kendall, T.J. Scar-associated macrophages are a major source of hepatic matrix metalloproteinase-13 and facilitate the resolution of murine hepatic fibrosis. *J. Immunol.* **2007**, *178*, 5288–5295. [CrossRef] [PubMed]

57. Ramachandran, P.; Pellicoro, A.; Vernon, M.A.; Boulter, L.; Aucott, R.L.; Ali, A.; Hartland, S.N.; Snowdon, V.K.; Cappon, A.; Gordon-Walker, T.T.; et al. Differential Ly-6C expression identifies the recruited macrophage phenotype, which orchestrates the regression of murine liver fibrosis. *Proc. Natl. Acad. Sci. USA* **2012**, *109*, E3186–E3195. [CrossRef] [PubMed]

58. Thomas, J.A.; Pope, C.; Wojtacha, D.; Robson, A.J.; Gordon-Walker, T.T.; Hartland, S.; Ramachandran, P.; Van Deemter, M.; Hume, D.A.; Iredale, J.P.; et al. Macrophage Therapy for Murine Liver Fibrosis Recruits Host Effector Cells Improving Fibrosis, Regeneration, and Function. *Hepatology* **2011**, *53*, 2003–2015. [CrossRef]

59. Thomson, J.A.; Itskovitz-Eldor, J.; Shapiro, S.S.; Waknitz, M.A.; Swiergiel, J.J.; Marshall, V.S.; Jones, J.M. Embryonic stem cell lines derived from human blastocysts. *Science* **1998**, *282*, 1145–1147. [CrossRef]

60. Reubinoff, B.E.; Pera, M.F.; Fong, C.Y.; Trounson, A.; Bongso, A. Embryonic stem cell lines from human blastocysts: somatic differentiation in vitro. *Nat. Biotechnol.* **2000**, *18*, 399–404. [CrossRef]

61. Brolen, G.; Sivertsson, L.; Bjorquist, P.; Eriksson, G.; Ek, M.; Semb, H.; Johansson, I.; Andersson, T.B.; Ingelman-Sundberg, M.; Heins, N. Hepatocyte-like cells derived from human embryonic stem cells specifically via definitive endoderm and a progenitor stage. *J. Biotechnol.* **2010**, *145*, 284–294. [CrossRef]

62. Hay, D.C.; Fletcher, J.; Payne, C.; Terrace, J.D.; Gallagher, R.C.J.; Snoeys, J.; Black, J.R.; Wojtacha, D.; Samuel, K.; Hannoun, Z.; et al. Highly efficient differentiation of hESCs to functional hepatic endoderm requires ActivinA and Wnt3a signaling. *Proc. Natl. Acad. Sci. USA* **2008**, *105*, 12301–12306. [CrossRef] [PubMed]

63. Woo, D.H.; Kim, S.K.; Lim, H.J.; Heo, J.; Park, H.S.; Kang, G.Y.; Kim, S.E.; You, H.J.; Hoeppner, D.J.; Kim, Y.; et al. Direct and Indirect Contribution of Human Embryonic Stem Cell-Derived Hepatocyte-Like Cells to Liver Repair in Mice. *Gastroenterology* **2012**, *142*, 602–611. [CrossRef] [PubMed]

64. Duan, Y.Y.; Catana, A.; Meng, Y.; Yamamoto, N.; He, S.Q.; Gupta, S.; Gambhir, S.S.; Zerna, M.A. Differentiation and enrichment of hepatocyte-like cells from human embryonic stem cells in vitro and in vivo. *Stem Cells* **2007**, *25*, 3058–3068. [CrossRef] [PubMed]

65. Lavon, N.; Yanuka, O.; Benvenisty, N. Differentiation and isolation of hepatic-like cells from human embryonic stem cells. *Differentiation* **2004**, *72*, 230–238. [CrossRef] [PubMed]

66. Basma, H.; Soto-Gutierrez, A.; Yannam, G.R.; Liu, L.P.; Ito, R.; Yamamoto, T.; Ellis, E.; Carson, S.D.; Sato, S.; Chen, Y.; et al. Differentiation and Transplantation of Human Embryonic Stem Cell-Derived Hepatocytes. *Gastroenterology* **2009**, *136*, 990–999. [CrossRef]

67. Yamamoto, H.; Quinn, G.; Asari, A.; Yamanokuchi, H.; Teratani, T.; Terada, M.; Ochiya, T. Differentiation of embryonic stem cells into hepatocytes: Biological functions and therapeutic application. *Hepatology* **2003**, *37*, 983–993. [CrossRef] [PubMed]

68. Tolosa, L.; Caron, J.; Hannoun, Z.; Antoni, M.; Lopez, S.; Burks, D.; Castell, J.V.; Weber, A.; Gomez-Lechon, M.J.; Dubart-Kupperschmitt, A. Transplantation of hESC-derived hepatocytes protects mice from liver injury. *Stem Cell Res. Ther.* **2015**, *6*. [CrossRef]

69. Zhu, S.Y.; Rezvani, M.; Harbell, J.; Mattis, A.N.; Wolfe, A.R.; Benet, L.Z.; Willenbring, H.; Ding, S. Mouse liver repopulation with hepatocytes generated from human fibroblasts. *Nature* **2014**, *508*, 93–97. [CrossRef]

70. Takahashi, K.; Yamanaka, S. Induction of pluripotent stem cells from mouse embryonic and adult fibroblast cultures by defined factors. *Cell* **2006**, *126*, 663–676. [CrossRef]

71. Kia, R.; Sison, R.L.; Heslop, J.; Kitteringham, N.R.; Hanley, N.; Mills, J.S.; Park, B.K.; Goldring, C.E. Stem cell-derived hepatocytes as a predictive model for drug-induced liver injury: are we there yet? *Br. J. Clin. Pharmacol.* **2013**, *75*, 885–896. [CrossRef] [PubMed]

72. Gomez-Lechon, M.J.; Tolosa, L. Human hepatocytes derived from pluripotent stem cells: a promising cell model for drug hepatotoxicity screening. *Arch. Toxicol.* **2016**, *90*, 2049–2061. [CrossRef] [PubMed]

73. Zhao, T.B.; Zhang, Z.N.; Rong, Z.L.; Xu, Y. Immunogenicity of induced pluripotent stem cells. *Nature* **2011**, *474*, 212–215. [CrossRef] [PubMed]

74. Tolosa, L.; Pareja, E.; Gomez-Lechon, M. Clinical Application of Pluripotent Stem Cells: An Alternative Cell-Based Therapy for Treating Liver Diseases? *Transplantation* **2016**, *100*, 2548–2557. [CrossRef] [PubMed]

75. Araki, R.; Uda, M.; Hoki, Y.; Sunayama, M.; Nakamura, M.; Ando, S.; Sugiura, M.; Ideno, H.; Shimada, A.; Nifuji, A.; et al. Negligible immunogenicity of terminally differentiated cells derived from induced pluripotent or embryonic stem cells. *Nature* **2013**, *494*, 100. [CrossRef] [PubMed]

76. Forbes, S.J.; Newsome, P.N. New horizons for stem cell therapy in liver disease. *J. Hepatol.* **2012**, *56*, 496–499. [CrossRef] [PubMed]

77. Zhan, Y.; Wang, Y.; Wei, L.; Chen, H.; Cong, X.; Fei, R.; Gao, Y.; Liu, F. Differentiation of hematopoietic stem cells into hepatocytes in liver fibrosis in rats. *Transplant. Proc.* **2006**, *38*, 3082–3085. [CrossRef] [PubMed]

78. Terai, S.; Takami, T.; Yamamoto, N.; Fujisawa, K.; Ishikawa, T.; Urata, Y.; Tanimoto, H.; Iwamoto, T.; Mizunaga, Y.; Matsuda, T.; et al. Status and Prospects of Liver Cirrhosis Treatment by Using Bone Marrow-Derived Cells and Mesenchymal Cells. *Tissue Eng. Part B Rev.* **2014**, *20*, 206–210. [CrossRef]

79. Nakamura, T.; Torimura, T.; Sakamoto, M.; Hashimoto, O.; Taniguchi, E.; Inoue, K.; Sakata, R.; Kumashiro, R.; Murohara, T.; Ueno, T.; et al. Significance and therapeutic potential of endothelial progenitor cell transplantation in a cirrhotic liver rat model. *Gastroenterology* **2007**, *133*, 91–107. [CrossRef] [PubMed]

80. Dollé, L.; Best, J.; Mei, J.; Al Battah, F.; Reynaert, H.; van Grunsven, L.A.; Geerts, A. The quest for liver progenitor cells: A practical point of view. *J. Hepatol.* **2010**, *52*, 117–129. [CrossRef]

81. Rossi, L.; Challen, G.A.; Sirin, O.; Lin, K.K.-Y.; Goodell, M.A. Hematopoietic stem cell characterization and isolation. *Methods Mol. Biol.* **2011**, *750*, 47–59. [CrossRef]

82. Lagasse, E.; Connors, H.; Al-Dhalimy, M.; Reitsma, M.; Dohse, M.; Osborne, L.; Wang, X.; Finegold, M.; Weissman, I.L.; Grompe, M. Purified hematopoietic stem cells can differentiate into hepatocytes in vivo. *Nat. Med.* **2000**, *6*, 1229–1234. [CrossRef] [PubMed]

83. Yannaki, E.; Athanasiou, E.; Xagorari, A.; Constantinou, V.; Batsis, L.; Kaloyannidis, P.; Proya, E.; Anagnostopoulos, A.; Fassas, A. G-CSF-primed hematopoietic stem cells or G-CSF per se accelerate recovery and improve survival after liver injury, predominantly by promoting endogenous repair programs. *Exp. Hematol.* **2005**, *33*, 108–119. [CrossRef] [PubMed]

84. Vainshtein, J.M.; Kabarriti, R.; Mehta, K.J.; Roy-Chowdhury, J.; Guha, C. Bone Marrow-Derived Stromal Cell Therapy in Cirrhosis: Clinical Evidence, Cellular Mechanisms, and Implications for the Treatment of Hepatocellular Carcinoma. *Int. J. Radiat. Oncol. Biol. Phys.* **2014**, *89*, 786–803. [CrossRef] [PubMed]

85. Austin, T.W.; Lagasse, E. Hepatic regeneration from hematopoietic stem cells. *Mech. Dev.* **2003**, *120*, 131–135. [CrossRef]

86. Thorgeirsson, S.S.; Grisham, J.W. Hematopoietic cells as hepatocyte stem cells: A critical review of the evidence. *Hepatology* **2006**, *43*, 2–8. [CrossRef] [PubMed]

87. Larrivée, B.; Karsan, A. Involvement of Marrow-Derived Endothelial Cells in Vascularization. In *Bone Marrow-Derived Progenitors*; Handbook of Experimental Pharmacology; Kauser, K., Zeiher, A.M., Eds.; Springer: Berlin/Heidelberg, Germany, 2007.

88. Asahara, T.; Masuda, H.; Takahashi, T.; Kalka, C.; Pastore, C.; Silver, M.; Kearne, M.; Magner, M.; Isner, J.M. Bone marrow origin of endothelial progenitor cells responsible for postnatal vasculogenesis in physiological and pathological neovascularization. *Circ. Res.* **1999**, *85*, 221–228. [CrossRef] [PubMed]

89. Asahara, T.; Murohara, T.; Sullivan, A.; Silver, M.; vanderZee, R.; Li, T.; Witzenbichler, B.; Schatteman, G.; Isner, J.M. Isolation of putative progenitor endothelial cells for angiogenesis. *Science* **1997**, *275*, 964–967. [CrossRef]

90. Wang, L.; Wang, X.D.; Xie, G.H.; Hill, C.K.; DeLeve, L.D. Liver sinusoidal endothelial cell progenitor cells promote liver regeneration in rats. *J. Clin. Investig.* **2012**, *122*, 1567–1573. [CrossRef]

91. Taniguchi, E.; Kin, M.; Torimura, T.; Nakamura, T.; Kumemura, H.; Hanada, S.; Hisamoto, T.; Yoshida, T.; Kawaguchi, T.; Baba, S.; et al. Endothelial progenitor cell transplantation improves the survival following liver injury in mice. *Gastroenterology* **2006**, *130*, 521–531. [CrossRef]

92. Ueno, T.; Nakamura, T.; Torimura, T.; Sata, M. Angiogenic cell therapy for hepatic fibrosis. *Med. Mol. Morphol.* **2006**, *39*, 16–21. [CrossRef]

93. Dominici, M.; Le Blanc, K.; Mueller, I.; Slaper-Cortenbach, I.; Marini, F.C.; Krause, D.S.; Deans, R.J.; Keating, A.; Prockop, D.J.; Horwitz, E.M. Minimal criteria for defining multipotent mesenchymal stromal cells. The International Society for Cellular Therapy position statement. *Cytotherapy* **2006**, *8*, 315–317. [CrossRef] [PubMed]

94. Bianco, P.; Riminucci, M.; Gronthos, S.; Robey, P.G. Bone marrow stromal stem cells: Nature, biology, and potential applications. *Stem Cells* **2001**, *19*, 180–192. [CrossRef] [PubMed]

95. Rodriguez, A.M.; Elabd, C.; Amri, E.Z.; Ailhaud, G.; Dani, C. The human adipose tissue is a source of multipotent stem cells. *Biochimie* **2005**, *87*, 125–128. [CrossRef] [PubMed]

96. Araujo, A.B.; Furlan, J.M.; Salton, G.D.; Schmalfuss, T.; Rohsig, L.M.; Silla, L.M.R.; Passos, E.P.; Paz, A.H. Isolation of human mesenchymal stem cells from amnion, chorion, placental decidua and umbilical cord: comparison of four enzymatic protocols. *Biotechnol. Lett.* **2018**, *40*, 989–998. [CrossRef] [PubMed]

97. Burra, P.; Arcidiacono, D.; Bizzaro, D.; Chioato, T.; Di Liddo, R.; Banerjee, A.; Cappon, A.; Bo, P.; Conconi, M.T.; Parnigotto, P.P.; et al. Systemic administration of a novel human umbilical cord mesenchymal stem cells population accelerates the resolution of acute liver injury. *BMC Gastroenterol.* **2012**, *12*. [CrossRef] [PubMed]

98. Banerjee, A.; Bizzaro, D.; Burra, P.; Di Liddo, R.; Pathak, S.; Arcidiacono, D.; Cappon, A.; Bo, P.; Conconi, M.T.; Crescenzi, M.; et al. Umbilical cord mesenchymal stem cells modulate dextran sulfate sodium induced acute colitis in immunodeficient mice. *Stem Cell Res. Ther.* **2015**, *6*. [CrossRef]

99. Christ, B.; Bruckner, S.; Winkler, S. The Therapeutic Promise of Mesenchymal Stem Cells for Liver Restoration. *Trends Mol. Med.* **2015**, *21*, 673–686. [CrossRef] [PubMed]

100. Andrzejewska, A.; Lukomska, B.; Janowski, M. Concise Review: Mesenchymal Stem Cells: From Roots to Boost. *Stem Cells* **2019**, *37*. [CrossRef]

101. Huch, M.; Gehart, H.; van Boxtel, R.; Hamer, K.; Blokzijl, F.; Verstegen, M.M.A.; Ellis, E.; van Wenum, M.; Fuchs, S.A.; de Ligt, J.; et al. Long-Term Culture of Genome-Stable Bipotent Stem Cells from Adult Human Liver. *Cell* **2015**, *160*, 299–312. [CrossRef]

102. Lanzoni, G.; Cardinale, V.; Carpino, G. The Hepatic, Biliary, and Pancreatic Network of Stem/Progenitor Cell Niches in Humans: A New Reference Frame for Disease and Regeneration. *Hepatology* **2016**, *64*, 277–286. [CrossRef] [PubMed]

103. Schmelzer, E.; Zhang, L.; Bruce, A.; Wauthier, E.; Ludlow, J.; Yao, H.L.; Moss, N.; Melhem, A.; McClelland, R.; Turner, W.; et al. Human hepatic stem cells from fetal and postnatal donors. *J. Exp. Med.* **2007**, *204*, 1973–1987. [CrossRef] [PubMed]

104. Libbrecht, L. Hepatic progenitor cells in human liver tumor development. *World J. Gastroenterol.* **2006**, *12*, 6261–6265. [CrossRef] [PubMed]

105. Huch, M.; Dorrell, C.; Boj, S.F.; van Es, J.H.; Li, V.S.; van de Wetering, M.; Sato, T.; Hamer, K.; Sasaki, N.; Finegold, M.J.; et al. In vitro expansion of single Lgr5+ liver stem cells induced by Wnt-driven regeneration. *Nature* **2013**, *494*, 247–250. [CrossRef] [PubMed]

106. Huch, M.; Boj, S.F.; Clevers, H. Lgr5(+) liver stem cells, hepatic organoids and regenerative medicine. *Regen Med.* **2013**, *8*, 385–387. [CrossRef] [PubMed]

107. Takebe, T.; Sekine, K.; Enomura, M.; Koike, H.; Kimura, M.; Ogaeri, T.; Zhang, R.R.; Ueno, Y.; Zheng, Y.W.; Koike, N.; et al. Vascularized and functional human liver from an iPSC-derived organ bud transplant. *Nature* **2013**, *499*, 481–484. [CrossRef] [PubMed]

108. Willyard, C. The boom in mini stomachs, brains, breasts, kidneys and more. *Nat. News* **2015**, *523*, 520–522. [CrossRef] [PubMed]

109. Huch, M.; Koo, B.K. Modeling mouse and human development using organoid cultures. *Development* **2015**, *142*, 3113–3125. [CrossRef]

110. Shafiee, A.; Atala, A. Tissue Engineering: Toward a New Era of Medicine. *Annu. Rev. Med.* **2017**, *68*, 29–40. [CrossRef]

111. Ehrbar, M.; Sala, A.; Lienemann, P.; Ranga, A.; Mosiewicz, K.; Bittermann, A.; Rizzi, S.C.; Weber, F.E.; Lutolf, M.P. Elucidating the role of matrix stiffness in 3D cell migration and remodeling. *Biophys. J.* **2011**, *100*, 284–293. [CrossRef]

112. Esch, M.B.; Prot, J.M.; Wang, Y.I.; Miller, P.; Llamas-Vidales, J.R.; Naughton, B.A.; Applegate, D.R.; Shuler, M.L. Multi-cellular 3D human primary liver cell culture elevates metabolic activity under fluidic flow. *Lab Chip* **2015**, *15*, 2269–2277. [CrossRef] [PubMed]

113. Skardal, A.; Shupe, T.; Atala, A. Organoid-on-a-chip and body-on-a-chip systems for drug screening and disease modeling. *Drug Discov. Today* **2016**, *21*, 1399–1411. [CrossRef] [PubMed]

114. Chen, F.M.; Liu, X.H. Advancing biomaterials of human origin for tissue engineering. *Prog. Polym. Sci.* **2016**, *53*, 86–168. [CrossRef] [PubMed]

115. Maghsoudlou, P.; Georgiades, F.; Smith, H.; Milan, A.; Shangaris, P.; Urbani, L.; Loukogeorgakis, S.P.; Lombardi, B.; Mazza, G.; Hagen, C.; et al. Optimization of Liver Decellularization Maintains Extracellular Matrix Micro-Architecture and Composition Predisposing to Effective Cell Seeding. *PLoS ONE* **2016**, *11*. [CrossRef]

116. Andrews, E. Damage of Porcine Aortic-Valve Tissue Caused by the Surfactant Sodiumdodecylsulphate. *Thorac. Cardiovasc. Surg.* **1986**, *34*, 340–341. [CrossRef]

117. Ott, H.C.; Matthiesen, T.S.; Goh, S.K.; Black, L.D.; Kren, S.M.; Netoff, T.I.; Taylor, D.A. Perfusion-decellularized matrix: using nature's platform to engineer a bioartificial heart. *Nat. Med.* **2008**, *14*, 213–221. [CrossRef] [PubMed]

118. Uygun, B.E.; Soto-Gutierrez, A.; Yagi, H.; Izamis, M.L.; Guzzardi, M.A.; Shulman, C.; Milwid, J.; Kobayashi, N.; Tilles, A.; Berthiaume, F.; et al. Organ reengineering through development of a transplantable recellularized liver graft using decellularized liver matrix. *Nat. Med.* **2010**, *16*, 814–820. [CrossRef] [PubMed]

119. Kadota, Y.; Yagi, H.; Inomata, K.; Matsubara, K.; Hibi, T.; Abe, Y.; Kitago, M.; Shinoda, M.; Obara, H.; Itano, O.; et al. Mesenchymal stem cells support hepatocyte function in engineered liver grafts. *Organogenesis* **2014**, *10*, 268–277. [CrossRef]

120. Park, K.M.; Hussein, K.H.; Hong, S.H.; Ahn, C.; Yang, S.R.; Park, S.M.; Kweon, O.K.; Kim, B.M.; Woo, H.M. Decellularized Liver Extracellular Matrix as Promising Tools for Transplantable Bioengineered Liver Promotes Hepatic Lineage Commitments of Induced Pluripotent Stem Cells. *Tissue Eng. Part A* **2016**, *22*, 449–460. [CrossRef]

121. Ogiso, S.; Yasuchika, K.; Fukumitsu, K.; Ishii, T.; Kojima, H.; Miyauchi, Y.; Yamaoka, R.; Komori, J.; Katayama, H.; Kawai, T.; et al. Efficient recellularisation of decellularised whole-liver grafts using biliary tree and foetal hepatocytes. *Sci. Rep.* **2016**, *6*, 35887. [CrossRef]

122. Devalliere, J.; Chen, Y.B.; Dooley, K.; Yarmush, M.L.; Uygun, B.E. Improving functional re-endothelialization of acellular liver scaffold using REDV cell-binding domain. *Acta Biomater.* **2018**, *78*, 151–164. [CrossRef]

123. Wu, Q.; Bao, J.; Zhou, Y.J.; Wang, Y.J.; Du, Z.G.; Shi, Y.J.; Li, L.; Bu, H. Optimizing Perfusion-Decellularization Methods of Porcine Livers for Clinical-Scale Whole-Organ Bioengineering. *BioMed Res. Int.* **2015**. [CrossRef]

124. Verstegen, M.M.A.; Willemse, J.; van den Hoek, S.; Kremers, G.J.; Luider, T.M.; van Huizen, N.A.; Willemssen, F.; Metselaar, H.J.; Ijzermans, J.N.M.; van der Laan, L.J.W.; et al. Decellularization of Whole Human Liver Grafts Using Controlled Perfusion for Transplantable Organ Bioscaffolds. *Stem Cells Dev.* **2017**, *26*, 1304–1315. [CrossRef]

125. Mattei, G.; Magliaro, C.; Pirone, A.; Ahluwalia, A. Decellularized Human Liver Is Too Heterogeneous for Designing a Generic Extracellular Matrix Mimic Hepatic Scaffold. *Artif. Organs* **2017**, *41*, E347–E355. [CrossRef]

126. Turner, W.S.; Schmelzer, E.; McClelland, R.; Wauthier, E.; Chen, W.; Reid, L.M. Human hepatoblast phenotype maintained by hyaluronan hydrogels. *J. Biomed. Mater. Res. Part B Appl. Biomater.* **2007**, *82B*, 156–168. [CrossRef]

127. Richert, L.; Binda, D.; Hamilton, G.; Viollon-Abadie, C.; Alexandre, E.; Bigot-Lasserre, D.; Bars, R.; Coassolo, P.; LeCluyse, E. Evaluation of the effect of culture configuration on morphology, survival time, antioxidant status and metabolic capacities of cultured rat hepatocytes. *Toxicol. In Vitro* **2002**, *16*, 89–99. [CrossRef]

128. Fu, R.H.; Wang, Y.C.; Liu, S.P.; Huang, C.M.; Kang, Y.H.; Tsai, C.H.; Shyu, W.C.; Lin, S.Z. Differentiation of Stem Cells: Strategies for Modifying Surface Biomaterials. *Cell Transplant.* **2011**, *20*, 37–47. [CrossRef]

129. Kleinman, H.K.; Martin, G.R. Matrigel: basement membrane matrix with biological activity. *Semin Cancer Biol.* **2005**, *15*, 378–386. [CrossRef]

130. Jain, E.; Damania, A.; Kumar, A. Biomaterials for liver tissue engineering. *Hepatol. Int.* **2014**, *8*, 185–197. [CrossRef]

131. Wang, H.P.; Shi, Q.; Guo, Y.N.; Li, Y.N.; Sun, T.; Huang, Q.; Fukuda, T. Contact assembly of cell-laden hollow microtubes through automated micromanipulator tip locating. *J. Micromech. Microeng.* **2017**, *27*. [CrossRef]

132. Jammalamadaka, U.; Tappa, K. Recent Advances in Biomaterials for 3D Printing and Tissue Engineering. *J. Funct. Biomater.* **2018**, *9*. [CrossRef]

133. Hospodiuk, M.; Dey, M.; Sosnoski, D.; Ozbolat, I.T. The bioink: A comprehensive review on bioprintable materials. *Biotechnol. Adv.* **2017**, *35*, 217–239. [CrossRef]

134. Liu, W.J.; Zhang, Y.S.; Heinrich, M.A.; De Ferrari, F.; Jang, H.L.; Bakht, S.M.; Alvarez, M.M.; Yang, J.Z.; Li, Y.C.; Trujillo-de Santiago, G.; et al. Rapid Continuous Multimaterial Extrusion Bioprinting. *Adv. Mater.* **2017**, *29*. [CrossRef]

135. Murphy, S.V.; Atala, A. 3D bioprinting of tissues and organs. *Nat. Biotechnol.* **2014**, *32*, 773–785. [CrossRef]

136. Murphy, S.V.; Skardal, A.; Atala, A. Evaluation of hydrogels for bio-printing applications. *J. Biomed. Mater. Res. Part A* **2013**, *101*, 272–284. [CrossRef]

137. Skardal, A.; Atala, A. Biomaterials for Integration with 3-D Bioprinting. *Ann. Biomed. Eng.* **2015**, *43*, 730–746. [CrossRef]

138. Arslan-Yildiz, A.; El Assal, R.; Chen, P.; Guven, S.; Inci, F.; Demirci, U. Towards artificial tissue models: past, present, and future of 3D bioprinting. *Biofabrication* **2016**, *8*. [CrossRef]

139. Li, X.; He, J.K.; Liu, Y.X.; Zhao, Q.; Wu, W.Q.; Li, D.C.; Jin, Z.M. Biomaterial Scaffolds with Biomimetic Fluidic Channels for Hepatocyte Culture. *J. Bionic Eng.* **2013**, *10*, 57–64. [CrossRef]

140. Griffith, L.G.; Swartz, M.A. Capturing complex 3D tissue physiology in vitro. *Nat. Rev. Mol. Cell Biol.* **2006**, *7*, 211–224. [CrossRef] [PubMed]

141. Zavan, B.; Brun, P.; Vindigni, V.; Amadori, A.; Habeler, W.; Pontisso, P.; Montemurro, D.; Abatangelo, G.; Cortivo, R. Extracellular matrix-enriched polymeric scaffolds as a substrate for hepatocyte cultures: in vitro and in vivo studies. *Biomaterials* **2005**, *26*, 7038–7045. [CrossRef] [PubMed]

142. Gerlach, J.C.; Encke, J.; Hole, O.; Muller, C.; Ryan, C.J.; Neuhaus, P. BIOREACTOR FOR A LARGER SCALE HEPATOCYTE IN-VITRO PERFUSION. *Transplantation* **1994**, *58*, 984–988. [CrossRef]

143. Lee, M.Y.; Kumar, R.A.; Sukumaran, S.M.; Hogg, M.G.; Clark, D.S.; Dordick, J.S. Three-dimensional cellular microarray for high-throughput toxicology assays. *Proc. Natl. Acad. Sci. USA* **2008**, *105*, 59–63. [CrossRef]

144. Andria, B.; Bracco, A.; Cirino, G.; Chamuleau, R.A.F.M. Liver Cell Culture Devices. *Cell Med.* **2010**, *1*, 55–70. [CrossRef]

145. Allen, J.W.; Khetani, S.R.; Bhatia, S.N. In vitro zonation and toxicity in a hepatocyte bioreactor. *Toxicol. Sci.* **2005**, *84*, 110–119. [CrossRef]

146. Lee-Montiel, F.T.; George, S.M.; Gough, A.H.; Sharma, A.D.; Wu, J.F.; DeBiasio, R.; Vernetti, L.A.; Taylor, D.L. Control of oxygen tension recapitulates zone-specific functions in human liver microphysiology systems. *Exp. Biol. Med.* **2017**, *242*, 1617–1632. [CrossRef]

147. Goulet, F.; Normand, C.; Morin, O. Cellular interactions promote tissue-specific function, biomatrix deposition and junctional communication of primary cultured hepatocytes. *Hepatology* **1988**, *8*, 1010–1018. [CrossRef]

148. Baudoin, R.; Corlu, A.; Griscom, L.; Legallais, C.; Leclerc, E. Trends in the development of microfluidic cell biochips for in vitro hepatotoxicity. *Toxicol. In Vitro* **2007**, *21*, 535–544. [CrossRef]

149. Banaeiyan, A.A.; Theobald, J.; Paukstyte, J.; Wolfl, S.; Adiels, C.B.; Goksor, M. Design and fabrication of a scalable liver-lobule-on-a-chip microphysiological platform. *Biofabrication* **2017**, *9*, 015014. [CrossRef]

bioengineering

MDPI

Article

Oxygen Persufflation in Liver Transplantation Results of a Randomized Controlled Trial

Anja Gallinat [1], Dieter Paul Hoyer [1], Georgios Sotiropoulos [1], Jürgen Treckmann [1], Tamas Benkoe [1], Jennifer Belker [1], Fuat Saner [1], Andreas Paul [1] and Thomas Minor [2,*]

[1] General, Visceral and Transplantation Surgery, University Hospital Essen, University Essen-Duisburg, 45147 Essen, Germany; anja.gallinat@uk-essen.de (A.G.); dieter.hoyer@uk-essen.de (D.P.H.); georgios.sotiropoulos@uk-essen.de (G.S.); Juergen-Walter.treckmann@uk-essen.de (J.T.); tamas.benkoe@uk-essen.de (T.B.); jennifer.belker@uk-essen.de (J.B.); fuat.saner@uk-essen.de (F.S.); andreas.paul@uk-essen.de (A.P.)

[2] Surgical Research Department, General, Visceral and Transplantation Surgery, University Hospital Essen, University Essen-Duisburg, 45147 Essen, Germany

* Correspondence: thomas.minor@uk-essen.de; Tel.: +49-(0)201-723-2713

Received: 14 March 2019; Accepted: 25 April 2019; Published: 27 April 2019

Abstract: Oxygen persufflation has shown experimentally to favorably influence hepatic energy dependent pathways and to improve survival after transplantation. The present trial evaluated oxygen persufflation as adjunct in clinical liver preservation. A total of $n = 116$ adult patients (age: 54 (23–68) years, M/F: 70/46), were enrolled in this prospective randomized study. Grafts were randomized to either oxygen persufflation for $\geq$2 h (O2) or mere cold storage (control). Only liver grafts from donors $\geq$55 years and/or marginal grafts after multiple rejections by other centers were included. Primary endpoint was peak-aspartate aminotransferase (AST) level until post-operative day 3. Standard parameters including graft- and patient survival were analyzed by uni- and multivariate analysis. Both study groups were comparable except for a longer ICU stay (4 versus 3 days) of the donors and a higher recipient age (57 versus 52 years) in the O2-group. Serum levels of TNF alpha were significantly reduced after oxygen persufflation ($p < 0.05$). Median peak-AST values did not differ between the groups (O2: 580 U/L, control: 699 U/L). Five year graft- and patient survival was similar. Subgroup analysis demonstrated a positive effect of oxygen persufflation concerning the development of early allograft dysfunction (EAD), in donors with a history of cardiopulmonary resuscitation and elevated ALT values, and concerning older or macrosteatotic livers. This study favors pre-implantation O2-persufflation in concrete subcategories of less than optimal liver grafts, for which oxygen persufflation can be considered a safe, cheap and easy applicable reconditioning method.

Keywords: liver transplantation; oxygen persufflation; reconditioning; randomized controlled trail

1. Introduction

The worldwide growing organ shortage led to an adjustment of thresholds to accept organs for transplantation. In the field of liver transplantation the numbers of donors aging > 65 years has increased more than ten-fold from 1991 to 2001 in the United Network for Organ sharing as well as the European Liver Transplant registry [1]. Likewise acceptance of steatotic organs and other risk afflicted organs has been increasing. Former studies demonstrated inferior outcomes for such organs [2–4]. Organs derived from suboptimal donors carry higher susceptibility of preservation/ischemia and reperfusion injuries, which ultimately results in higher rates of early allograft dysfunction after liver transplantation, a complication associated with reduced graft and patient survival [5,6].

Therefore, organ preservation techniques need to be adapted, contributing to the demands of suboptimal grafts. Optimized preservation techniques present a valuable opportunity to decrease ischemic organ injury and increase the number of viable donor organs and advance the total pool of organ donors.

The attenuation of ischemic organ damage can be elegantly achieved by the provision of gaseous oxygen during static cold storage. Development of preservation damage most likely depends on adequate redox and intracellular signal homeostasis. Venous oxygen persufflation [7] is thought to replenish depleted cellular energy stores in a simple and applicable way. As the majority of preservation/reperfusion injury arises at the time of the warm reperfusion of the organ [8,9], an end-ischemic adaption of the preservation method carries the possibility to prime the organs for this critical period of the transplantation process. Indeed, the optimal treatment time for a hypothermic reconditioning of cold stored liver grafts by gaseous oxygen persufflation was previously evaluated in a large animal model, demonstrating the best results after 2 h of end-ischemic reconditioning [10,11]. The mechanism includes the stabilization of cell and organ integrity by reduction of ischemia induced failure of cellular autophagy [12], which leads to an increased regenerative potential of the cells during the reperfusion period to clear impaired cell organelles and reprocess denatured proteins [13,14].

First clinical applications of this end-ischemic organ persufflation demonstrated feasibility and safety in the clinical setting [15]. Subsequently, Khorsandi and coworkers [16] were able to confirm gaseous oxygen insufflation to improve hepatic energy homeostasis at the end of ischemic preservation in human discard livers.

Based on these encouraging observations, a randomized controlled trial was created to systematically address, whether 2 h of gaseous oxygen persufflation of the isolated liver graft immediately prior to transplantation will improve early graft function upon reperfusion and mitigate adverse effects associated with preservation/reperfusion injury as compared to standard cold storage.

2. Patients and Methods

2.1. Study Design

The present study was carried out at a single center. It was designed as randomized, controlled, single blinded clinical study and comprised two arms (treatment versus control) (ISRCTN00167887). The study was approved by the local ethics committee (Ethics committee University Hospital Essen, AZ 09-4281) and followed the Declaration of Helsinki. The study was supported by the German Research Foundation (DFG MI 470/14/2).

2.2. Study Population

Only patients who met all inclusion and no exclusion criteria were included in the trial.

Inclusion criteria for the allografts were met when the organ was allocated by the "rescue offer" mechanism by EUROTRANSPLANT (see below) or when donor age was 55 years or older. (In contrast to the originally foreseen donor age criterion of >65 years, the required donor age has been reduced from 65 years to 55 years in order to cope with unexpectedly low numbers of organ offers at the time of enrollment and to safeguard the timely completion of the study. A respective official amendment been approved by the local authorities.)

For the inclusion criteria of the liver transplant recipients the following requirements had to be met:

- Adult patients (>18 years of age)
- Recipients undergoing the first liver transplantation
- Willingness and ability to attend regular follow up examinations
- Written informed consent

Exclusion criteria for the recipients included high urgency listing, participation in other clinical trials and positivity for HIV.

2.3. Study Procedures

After acceptance of organ offers all livers were inspected at the local transplant center by an experienced transplant surgeon. Liver zero-biopsies were done by the acting implant surgeon, whenever the macroscopic appearance of the liver deemed questionable. Thus, in some cases, livers were transplanted without prior histology.

When found to be suitable for transplantation randomization was initiated after verification of all inclusion and exclusion criteria. Randomization was technically realized by a web-interface organized by the center for clinical trials Essen with 1:1 randomization ratio as per computer-generated randomization schedule. Variable block sizes were used with patient level stratification for Model for End-Stage Liver Disease (MELD) score (3 groups: <20, 20–30, and >30).

After randomization to one of the study arms all procedures strictly followed the study protocol: For the treatment group, donor livers were subjected to 2 h of venous systemic oxygen persufflation (OPAL) as detailed previously [15,17]. Shortly, allografts were stored in ice-cold preservation solution during the procedure and backtable preparation was carried out as usual. Additionally, a catheter was inserted into the suprahepatic vena cava. An atraumatic clamp temporarily closed the infrahepatic vena cava. Filtered (membrane pore size of 5 μm) and humidified oxygen gas was then introduced via the catheter in the suprahepatic caval vein at a pressure limited to 18 mmHg to avoid barotraumata of the vasculature. An endosufflator (WISAP GmbH Sauerlach, Germany), which was technically modified for the use of oxygen instead of carbon dioxide, was utilized. When persufflation is started postsinusoidal venules become dilated due to the pressure applied to the hepatic venous system and gas bubbles rise up via the portal vain. Additionally, small pinpricks are set with a 27 gauge needle into dilated venules at the periphery of the liver lobes that also allow the oxygen to leave the microvasculature [17,18]. In contrast to the machine perfusion technologies, no liquid perfusion is involved in the oxygenation of the liver tissue, that takes place exclusively by gaseous diffusion. The persufflation method hence does not require the use of additional oxygenators or disposable perfusion kits.

For the control group livers were kept simply cold stored until implantation after usual backtable procedures.

After transplantation all patients were observed for seven days on a daily basis. Additional follow up visits were carried out on the day of discharge, 3, 6, and 12 months after transplantation. Further follow up included common visits at our outpatient clinic.

2.4. Objectives and Endpoints

Primary endpoint was the peak value of serum aspartate aminotransferase (AST) during the first three days after liver transplantation. Secondary endpoints were graft and patient survival, rate of re-transplantation, early allograft function (EAD, see below), ICU stay, time of postoperative ventilation and dialysis as well as morbidity according to Dindo-Clavien Classification Grade ≥ 3. Moreover serum levels of TNF-alpha were determined to investigate the impact of hypothermic reconditioning on pro-inflammatory upregulation after reperfusion. Serum samples taken one hour after reperfusion were analyzed using commercial ELISA kits on a fluorescence micro plate reader (Tecan, Grailsheim, Germany) according to the instructions of the manufacturer (R&D Systems, Wiesbaden, Germany).

2.5. Surgical Procedure and Immunosuppression

All organ procurements were carried out by specialized local teams according to the standards of the local procurement organizations within the different EUROTRANSPLANT regions. Orthotopic liver transplantation was performed with vena cava replacement and end-to-end-anastomosis of portal vein, hepatic artery and bile duct. All patients were treated at the ICU after transplantation. The perioperative care was similar in both groups as well as the concept of immunosuppression. Intravenous corticosteroids (1000 mg methylprednisolone) were applied intraoperatively. Postoperatively, tacrolimus (adjusted in accordance to the trough level of the drug) in combination with corticosteroids and mycophenolate mofetil were utilized.

2.6. Definition of Rescue Allocation

Livers refused by more than three different centers for allocated candidates with the highest MELD scores on the national waiting list were characterized as "organ rescue offers". These grafts were then either offered to the nearest center with a suitable recipient or allocated to the first center to accept them (multiple-refusal/competitive rescue offer procedure). "Organ rescue offers" were also occasionally encountered in instances of donor instability, prolonged cold ischemic times, or unfavorable logistic reasons.

2.7. Definition of Early Allograft Dysfunction (EAD)

Early Allograft Dysfunction (EAD) was defined as: Bilirubin ≥ 10 mg/dL on postoperative day 7 and/or INR ≥ 1.6 on postoperative day 7 and/or AST or ALT > 2000 IU/L within the first 7 days [19]. Each case was classified as "EAD" or "no-EAD."

2.8. Clinical Factors for Outcome Analysis

The following donor factors: age, gender, BMI, cause of death (cerebrovascular accident, hypoxia, trauma, others), cold ischemic time, ICU length of stay, biopsy proven steatosis (macrovesicular and microvesicular), organ protection solution used during the procurement (HTK, UW), last laboratory values (AST, ALT, gGT, Bilirubin, Creatinine, Serum Sodium, INR) and the Donor Risk Index (DRI) [20]—for the calculation of the DRI "race" was always set to "Caucasian". The following recipient factors were analyzed: age, gender, BMI, etiology of liver disease, laboratory Model for End stage Liver Disease (MELD) score before transplantation, time for surgical procedure, warm ischemic time, hospitality stay, and rejection within 3 month. Charlson Co-morbidity index was calculated according to Charlson et al. [21].

2.9. Monitoring, Data Safety Monitoring Board

All trial related procedures were monitored and controlled by the center for clinical trials Essen (ZKSE), according to ICH-GCP guidelines. Additionally, an independent data safety monitoring board (DSMB), consisting of a clinician, scientist and statistician closely followed the proper conduct of the trial and all severe adverse events (SAE). SAE were defined as life threatening or deadly events or events that entail permanent injuries or require prolongation of hospital stay. None of the occurring complications (the frequency of which did not differ between the groups) gave reason for intervention by the safety board.

2.10. Sample Size Calculation and Statistical Analysis

The sample size calculation was performed to detect a relative effect of $p = 0.66$ (which is comparable to a mean difference of ~0.6 in units of standard deviations of a standard normal distribution) for the primary endpoint (maximum AST value during the first three days after transplantation) with a power of 0.8 when significance is set to a = 0.05 (two-sided). A drop-out rate of 10% was assumed, resulting in a sample size of 58 patients per groups (total 116 patients).

Data were expressed as mean and standard error of the mean or median and range values, as appropriate. Categorical variables were analyzed by chi-squared test. Continuous variables were analyzed by the Student t test or the Mann–Whitney U test. Treatment groups and clinical parameters were linked to the development of EAD after transplantation by univariable and multivariable logistic regression analysis censored for treatment groups. Factors with a p-value < 0.1 in either group were introduced into the respective multivariable model. Patient survival was calculated using the Kaplan–Meier method and compared with the log-rank test. Univariable and multivariable Cox

Proportional Hazard Analyses were carried out to delineate independent predictors of patient survival. Long-term patient survival was censored for patient death and treatment arm in order to investigate the impact of study procedures on the early outcome after transplantation and intervention. $p < 0.05$ was considered to be statistically significant. Statistical analyses were performed using JMP (version 10.0.0 SAS, SAS Institute Inc., Cary, NC, USA) and SPSS (IBM SPSS Statistics 24, IBM®, Armonk, NY, USA).

3. Results

3.1. Recruitment and Follow Up

Patients were enrolled and transplanted from 09/2011 to 12/2013. According to the study protocol 116 patients were recruited. Of these, 57 patients were randomized into the treatment group and 59 patients randomized into the control group. One year follow up was 100%. Graft and patient survival were analyzed for a total of 5 years after transplantation.

Only one patient was lost to follow up with a functioning allograft after more than 2 years after transplantation.

Median follow up time was 1466 days (1–2028 days). Enrollment of patients is shown in Figure 1.

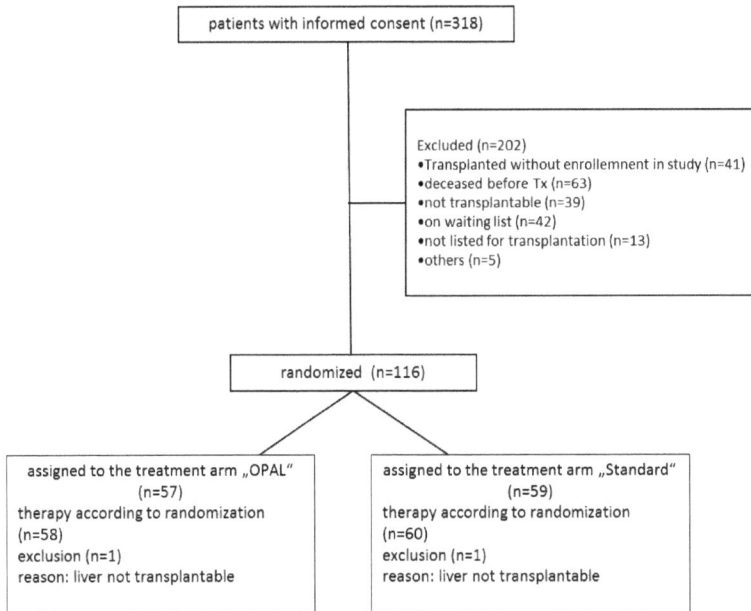

Figure 1. CONSORT diagram illustrating the study enrollment.

Following randomization 2 patients (1 in each treatment arm) were excluded as the donor organ has been considered inacceptable for transplantation after inspection by the responsible surgeon.

3.2. Donor, Recipient, and Perioperative Characteristics

Mean age of donor organs was 63 (±1.26) years. Half (50%) of the donors were male. Median donor ICU treatment before organ procurement was 3.0 (1–19) days. The median DRI was 1.8 ± 0.3. Cold preservation was applied for 452 ± 13.4 min. The warm ischemic time during the surgical procedure was 30 ± 0.6 min.

Recipients had a mean age of 53.2 ± 0.8 years and were predominantly male (60.3%). Indications (or a combination of indications) for OLT included cirrhosis related to alcoholic cirrhosis (31.9%), viral hepatitis (26.7%), hepatocellular carcinoma (26.7%), NASH (6.9%) and others (25.9%). The mean labMELD before liver transplantation was 14.6 ± 0.6. Median duration of the surgical procedure was 257 (155–661) min.

Further details regarding donor, recipient and perioperative characteristics in both study arms are given in Table 1.

Table 1. Donor and recipient data; values given as median and range, where appropriate, in brackets.

Parameter	OPAL ($n = 57$)	Control ($n = 59$)	p-Value
Donor Age (years)	64 (30–95)	63 (28–84)	0.57
Donor Gender (m/f) (%)	56/44	44/56	0.27
Donor BMI (kg/m^2)	25(19–42)	26 (19–51)	0.38
Donor ICU stay (days)	3 (1–16)	4 (1–19)	0.02
Donor Cause of death (n)			
Cerebrovascular	37	39	
Hypoxia	10	13	0.8
Trauma	5	4	
Others	5	3	
Donor aspartate aminotransferase (AST) (U/L)	48 (9–501)	41 (9–607)	0.8
Donor ALT (U/L)	32 (6–956)	33 (6–282)	0.87
Donor γGT (U/L)	46 (7–381)	57 (6–416)	0.29
Donor Sodium (μmol/L)	149 (132–163)	149 (132–169)	0.71
Donor Creatinin (μmol/L)	80 (32–689)	81 (33–265)	0.87
Donor Bilirubin (μmol/L)	9 (3.4–30)	8.2 (2.7–564)	0.16
Donor INR	1.13 (0.88–3.50)	1.12 (0.87–5.60)	0.81
Donor Risk Index	1.83 (1.1–2.5)	1.80 (1.1–2.5)	0.55
Allograft Histology (n)	49	47	
Macrosteatosis (≥20%)	13	6	0.09
Microsteatosis (%)	50 (5–95)	40 (0–90)	0.36
Perfusion solution HTK/ UW (n)	53/4	52/7	0.37
Cold Ischemia Time (min)	443 (289–1090)	390 (259–740)	0.12
Recipient Age (years)	57 (31/69)	52 (24–67)	0.046
Recipient Gender (m/f) (%)	38/19	32/27	0.17
Recipient BMI (kg/m^2)	27 (18–44)	25 (17–41)	0.14
Underlying disease (%)			
Viral Hepatitis	8	10	
HCC	14	16	
linebreak Cholestative disease	7	4	0.83
Alcohol	14	18	
NASH	3	3	
Others	11	8	
Charlson Co-morbidity Index	4 (1–8)	4 (2–8)	0.25
Laboratory Model for End-Stage Liver Disease (MELD)	13 (6–31)	15 (6–40)	0.16

In brief, clinically relevant characteristics were similar in both groups. Donor ICU stay was significantly shorter in the treatment group. Recipient age was significantly older in the treatment group. In both groups allocation was center based in 80%. Cold ischemia time was numerically longer in the treatment group ($p > 0.05$).

3.3. Surgical Study Procedures

Oxygen persufflation was applied for 137 (103–205) min in the treatment group. Treatment with persufflation did not result in any serious adverse event or allograft loss. Minor bleeding was observed from the pinpricks. These were not clinically relevant and stopped spontaneously or after minimal electrocoagulation. Median duration of the surgical procedure was 260 (176–460) min in the treatment arm and 250 (155–661) min in the control group. Warm ischemia time was similar, being 30 (16–41) min in the persufflation group and 29 (20–65) min in the control group. Statistical differences regarding surgical study procedures were not observed between groups.

Transfusion of packed red blood cells (PRBs) was low and the same in both groups: 16 patients (28%) in the treatment group were transfused with a median of 2 PRBs compared to 25 patients (42%) transfused with a median of 2 PRBs in the control group.

3.4. Primary Endpoint

Peak AST values within the first three days after liver transplantation were higher in the control group compared to the treatment group. However, this did not reach statistical significance (1246 (310–8064) versus 972 (194–17577), control versus OPAL; cf Figure 2).

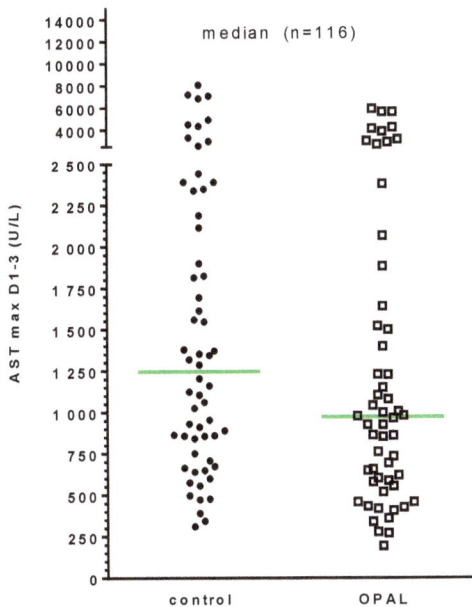

Figure 2. Peak values of AST during the first 3 days after transplantation.

3.5. Secondary Endpoints

For secondary endpoints several assessments of patient death and early graft function were compared between groups. Details are depicted in Table 2.

Rates of 30-day mortality and In-hospital mortality were not statistically different between groups. Few retransplantations were necessary in the present study: One patient developed primary non-function and died after retransplantation. Another patient in the treatment group developed arterial thrombosis one month after transplantation and was successfully retransplanted. This patient is now well and alive. Statistical comparison of PNF and retransplantation rates was not performed due to the low number of events. Early Allograft Dysfunction occurred in every fifth to fourth patient in both groups. Rate of postoperative acute kidney failure and the necessity for hemodialysis was higher in the control group, but did not reach statistical significance. Postoperative complications $\geq$ Grade III according to Dindo-Clavien were similar in both groups. Surrogates of complicated clinical courses like length of ICU stay and length of hospital stay did not show differences among the groups.

Table 2. Secondary outcome parameters: values given as median and range, where appropriate, in brackets.

Parameter	OPAL ($n = 57$)	Control ($n = 59$)	p-Value
Retransplantation (n)	2	-	-
Early allograft dysfunction (EAD) (n)	14	12	0.58
Recipient ICU stay (days)	3 (1–45)	3 (1–41)	0.97
Post Tx dialysis (n)	5	9	0.28
Recipient hospital stay (days)	20 (2–114)	18 (1–85)	0.07
30-day mortality (n)	3	2	0.62
In-hospital mortality (n)	5	5	0.95
Postop. Comlications (n)	22	17	
Dindo-Clavien IIIa	7	5	
Dindo-Clavien IIIb	8	3	0.26
Dindo-Clavien IVa	5	6	
Dindo-Clavien IVb	2	3	
Rejection within 3 months (n)	6	8	0.61

Serum levels of TNF alpha were found to be significantly reduced after oxygen persufflation: 11.1 ± 1.6 versus 5.9 ± 0.4 pg/ml; mean $\pm$ SEM, control versus OPAL $p < 0.05$.

3.6. Patient and Graft Outcome

Death censored graft survival in the treatment group was 89% after one, three, and five years. The control group demonstrated a graft survival of 87%, 84% and 82% after one, three, and five years, respectively ($p > 0.05$).

Patient survival in the complete study cohort was 80% and 70% after one and five years, respectively.

Patient survival in the treatment group was 77%, 74%, and 74% after one, three and five years. Accordingly, the control group demonstrated patient survival rates of 83%, 76%, and 66%. Comparison between groups demonstrated non-significant differences ($p = 0.56$).

Overall cause of death in the recipients after transplantation was in descending order: Tumor recurrence (10.4%), sepsis (8.6%), and HCV reinfection (6.9%).

3.7. Association of Clinical Parameters with Development of EAD

Logistic regression was performed to identify parameters associated with the development of EAD as a clinical marker for inferior allograft function. Results are displayed in Table 3.

For the treatment group, no clinical parameter was delineated as independent predictor for development of EAD after liver transplantation. In contrast, history of cardiopulmonary resuscitation and last donor ALT were significantly and independently associated with the development of EAD in the control group. Summarizing, oxygen persufflation showed a positive effect concerning the development of EAD in the case of donors with history of cardiopulmonary resuscitation and of donors with elevated ALT levels.

Table 3. Early allograft dysfunction; censored for treatment arm: Nominal logistic analysis, multivariate analysis and likelihood ratio test (*p*-values).

Parameter	OPAL		Control	
	Univariate	Multivariate	Univariate	Multivariate
Donor Age (years)	0.81		0.41	
Donor Age > 70years	0.7		0.65	
Donor BMI	0.09	3.26 0.07	0.67	0.14 0.71
Donor cause of death	0.31		0.23	
Donor ICU stay (days)	0.49		0.79	
Donor cardiopulmonary resuscitation	0.47	0.12 0.73	0.07	6.02 0.01
Donor AST (U/L)	0.87		0.1	
Donor ALT U/L)	0.38	2.54 0.11	0.03	5.25 0.02
Donor γGT (U/L)	0.73	0.97 0.32	0.03	1.25 0.26
Donor Bilirubin (μmol/L)	0.26		0.29	
Donor Risk Index	0.77		0.4	
Allograft histology: macrosteatosis	0.18		0.67	
Allograft histology: fibrosis	0.48		0.76	
Preservation solution	0.13	1.75 0.19	0.07	1.67 0.2
Cold Ischemia Time (h)	0.13		0.11	
Warm Ischemia Time (min)	0.61		0.66	
Duration of Surgical Procedure (min)	0.89		0.86	
Recipient Age (years)	0.36		0.6	
Recipient BMI	0.37	1.39 0.24	0.08	3.24 0.07
Lab-MELD score	0.52		0.43	

3.8. Association of Clinical Parameters with Patient Survival

Cox proportional hazard analysis was performed to identify parameters associated with long term patient survival (Table 4).

None of the analyzed factor demonstrated an independent association with the patient survival in the treatment group. On the contrary, two factors were significantly and independently associated with the patient survival in the control group: allograft macrosteatosis as a marker for graft quality and the MELD score of the recipients as a marker for severity of the underlying liver disease.

Quite of interest in this analysis was the strong tendency of donor age with the patient survival in the control group. Further analysis demonstrated that an advanced donor age of more than 70 years was significantly associated with the patient survival in univariable analysis. This was not observed in the treatment group. Due to the retroactive character of this analysis, donor age > 70 years was not introduced in the multivariable cox proportional model.

These findings could be interpreted as a positive effect of the revitalization treatment in the case of older livers (from donors > 70 years of age) or macrosteatotic livers, proposing selection criteria for the use of this method in the corresponding instances.

Table 4. Cox proportional hazard analysis (*p*-values), censored for long term patient survival and treatment arm.

Parameter	OPAL		Control	
	Univariate	Multivariate	Univariate	Multivariate
Donor age(years)	0.63	0.138 0.71	0.051	0.48 0.49
Donor age > 70 years	0.22		0.047	
Donor BMI	0.97		0.6	
Donor cause of death	0.17		0.99	
Donor ICU stay (days)	0.29		0.3	
Donor cardiopulmonary resuscitation	0.46		0.19	
Donor AST U/L)	0.87		0.53	
Donor ALT (U/L)	0.2		0.12	
Donor γGT (U/L)	0.67		0.69	
Donor Bilirubin μmol/L)	0.2		0.18	
Donor Risk Index	0.35		0.65	
Allograft histology: macrosteatosis	0.46	0.38 0.54	0.07	6.2 0.01
Allograft histology: fibrosis	0.09	1.17 0.28	0.44	3.69 0.06
Preservation solution	0.37		0.7	
Cold Ischemia Time (h)	0.55	0.05 0.83	0.09	3.55 0.06
Warm Ischemia Time (min)	0.83		0.27	
Duration of Surgical Procedure (min)	0.28		0.9	
Recipient Age (years)	0.71		0.17	
Recipient BMI	0.48		0.34	
MELD	0.49	0.48 0.49	0.054	5.06 0.03

3.9. Subgroup Analysis

The impact of advanced donor age was further investigated in a subgroup analysis. Patient survival was compared in recipients transplanted with organs from donors with age > 70 years to recipients transplanted with organs from donors with age < 70 years in the treatment group (*n* = 26) and control group (*n* = 23), respectively. Results are depicted in Figure 3.

In the treatment group patient survival was 80% after one and five years when transplantation was carried out with younger donors. In donors aged 70 or more the patient survival was 70% and 65% after one and five years, without statistical significant differences between groups. However, the same analysis in the control group demonstrated significantly worse survival after transplantation of allografts from donors aged 70 years or more with one and five year patient survival of 70% and 48%. Donors aged younger than 70 years led to a patient survival of 85% and 75% after one and five years, respectively.

We also evaluated maximal serum values of AST in the population receiving macrosteatotic livers (>20%) and found an accentuated trend towards a benefit in the OPAL group: 987 (271–3016) U/L versus 2498 (890–4332) U/L. However, because of the small number of patients (*n* = 12 in the OPAL group, *n* = 6 in the control group) differences were not evaluated for significance.

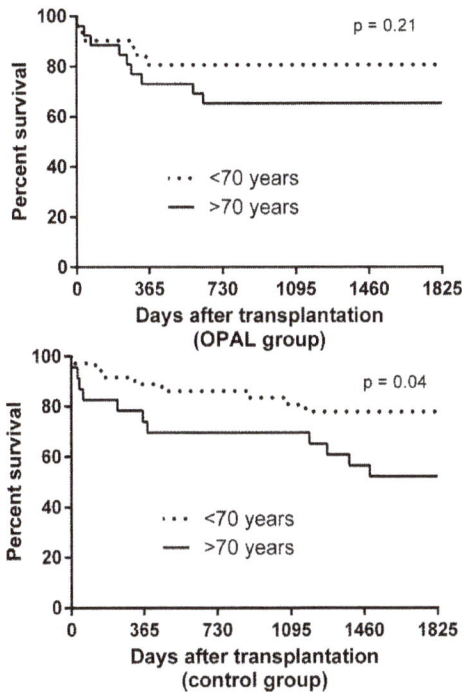

Figure 3. Five year patient survival in the treatment group (oxygen persufflation (OPAL)) and in the standard care group (control) according to recipient age (<70 years versus ≥70 years).

4. Discussion

This randomized controlled single center study investigated for the first time the impact of oxygen persufflation as adjunct in liver preservation on early allograft injury and dysfunction upon repefusion. The primary endpoint was aminotransferase peak of AST within the first three days after liver transplantation. While AST values were lower in the treatment group, statistical significance was not reached.

Assessment of secondary endpoints, such as early allograft dysfunction, primary non-function, and patient survival, showed a positive effect of the treatment on the development of EAD in the case of donors with history of cardiopulmonary resuscitation and of elevated ALT levels. Benefits with regard to patient survival were also present for marginal liver grafts with macrosteatosis or originating from donors aged > 70 years.

Furthermore a moderate, but significant reduction of TNF-alpha release could be documented for the entire collective.

Thus, a distinct clinical benefit for application of retrograde oxygen persufflation as reconditioning tool immediately before liver transplantation could be delineated in concrete subcategories of particularly endangered donor grafts. This comes to confirm the corresponding literature from experimental studies, as the concept of retrograde oxygen persufflation in liver allograft reconditioning is based on scientifically high grade research which has been published in the past decades: Initial reports by Isselhard [22] and Ross [23] demonstrated convincing results in animal kidneys already in the 1970s. Subsequently, the first clinical pilot project was successfully initiated in renal transplantation [24]. Experimental applications in liver allografts followed thereafter [25,26]. This research demonstrated significant reduction of non-parenchymal cell injury and vascular endothelial dysfunction after cold preservation of the liver by gaseous oxygen [27] as well as reduction of proteolysis leading to improved functional outcome after transplantation [28]. Gaseous oxygenation resulted in normalization of vascular resistance and reduced release of hepatocellular enzymes. Further investigations showed prevention of functional and ultrastructural impairments by venous oxygen persufflation [29] in steatotic rat livers.

In a porcine model, gaseous oxygen persufflation prevented primary non-function of livers after extended cold storage times and improved one week survival of the recipient from 0% up to 83% [11].

Based on these premises one might have expected a less equivocal result upon clinical usage oxygen persufflation. However, in contrast to the controlled experimental situation, a broader heterogeneity of included donor livers as well as recipient health status may influence the outcome after transplantation in the clinical setting.

Some of the donor organs might have had lesser needs for additional treatment than others, and, although distributed evenly across the groups, a notable fraction of not so marginal grafts might have obscured the benefit of oxygen persufflation in less resilient livers.

Such effect would not have been predictable, as extensive human studies are needed to delineate the specific influence of preservation techniques on each group of allografts. The subgroup analysis presented in this studies might be taken as hint that oxygen persufflation might be most effective in older donor allografts. While the study was not powered to prove this, the influence of donor age on patient survival in the control group, which was not traceable in the treatment group, suggests reduced damage to the allograft after oxygen persufflation.

In addition, results of multivariable studies demonstrated that history of donor cardiopulmonary resuscitation as well as elevated ALT levels in the donor contributed to development of EAD in the control group but not in the treatment group. This might indicate that after reconditioning of pre-damaged allografts, such risk factors lost their impact. The same was observed for patient survival: allograft macrosteatosis and MELD score were delineated as risk factors only in the control group. Against the background of similar patients and allografts transplanted in both groups, this resembles a surrogate of improved allografts after treatment by oxygen persufflation.

It should be of interest that the data at hand demonstrates excellent safety of the method applied. Serious adverse events related to treatment were not observed and all endpoints were qualitative similar to the common standard of cold storage.

Organ preservation is a science virtually lacking relevant clinical progress over more than 20 years.

Nowadays, several approaches of machine perfusion have been taken from experimental projects to first clinical applications [30–32]. These studies demonstrated feasibility and safety of such new methods and suggested beneficial effects on clinical outcomes like peak of aminotransferase after transplantation, development of EAD and maybe even biliary complications. Most importantly, these pilot projects led to the initiation of randomized controlled trials comparing these new methods with static cold storage. So far, no clear benefits in term of graft and patient survival could be demonstrated.

This again underscores the importance to identify more precisely those subgroups of allografts that need or may benefit from reconditioning measures and those who will not.

Compared with the more sophisticated methods of machine perfusion, the simple insufflation of gaseous oxygen excels by its ease of use and the unmatched cost-effectiveness may furthermore allow for a less critical application in attempt to rescue questionable liver grafts.

5. Conclusions

In conclusion, this randomized controlled trial demonstrated safety of venous oxygen persufflation of liver allografts immediately before transplantation as reconditioning tool. A clinical benefit could be demonstrated in concrete subcategories of less than optimal donor organs. Pending success of alternative new preservation methods might justify further clinical evaluation of oxygen persufflation as safe, cheap, and easy applicable reconditioning method in liver allografts subgroups.

Author Contributions: A.G. participated in performance of research, coordinated the collection of data and samples; A.G. and T.M. were responsible for the experimental therapy; D.P.H. and G.S. collected, analyzed and

Bioengineering **2019**, *6*, 35

interpreted data; J.T., T.B., F.S., and A.P. were responsible for the clinical conduct of the study; J.B. participated in data collection; A.P. and T.M. designed this study with help from other authors; All authors participated in the revision of the manuscript

Funding: The study was supported by the German Research Foundation (DFG MI 470/14/2).

Acknowledgments: We thank Mrs. Ose and her team for their continuous support, data monitoring and conducting some of the statistical analysis.

Conflicts of Interest: The authors declare no conflict of interest.

References

1. Nickkholgh, A.; Weitz, J.; Encke, J.; Sauer, P.; Mehrabi, A.; Büchler, M.W.; Schmidt, J.; Schemmer, P. Utilization of extended donor criteria in liver transplantation: a comprehensive review of the literature. *Nephrol. Dial. Transplant.* **2007**, *22*, viii29–viii36. [CrossRef] [PubMed]
2. Briceño, J.; Marchal, T.; Padillo, J.; Solórzano, G.; Pera, C. Influence of marginal donors on liver preservation injury. *Transplantation* **2002**, *74*, 522–526. [CrossRef]
3. Tector, A.J.; Mangus, R.S.; Chestovich, P.; Vianna, R.; Fridell, J.A.; Milgrom, M.L.; Sanders, C.; Kwo, P.Y. Use of Extended Criteria Livers Decreases Wait Time for Liver Transplantation Without Adversely Impacting Posttransplant Survival. *Ann. Surg.* **2006**, *244*, 439–450. [CrossRef] [PubMed]
4. Markmann, J.F.; Markmann, J.W.; Markmann, D.A.; Bacquerizo, A.; Singer, J.; Holt, C.D.; Gornbein, J.; Yersiz, H.; Morrissey, M.; Lerner, S.M.; et al. Preoperative factors associated with outcome and their impact on resource use in 1148 consecutive primary liver transplants. *Transplantation* **2001**, *72*, 1113–1122. [CrossRef] [PubMed]
5. Moore, D.E.; Feurer, I.D.; Speroff, T.; Gorden, D.L.; Wright, J.K.; Chari, R.S.; Pinson, C.W. Impact of Donor, Technical, and Recipient Risk Factors on Survival and Quality of Life After Liver Transplantation. *Arch. Surg.* **2005**, *140*, 273. [CrossRef]
6. Hoyer, D.P.; Paul, A.; Gallinat, A.; Molmenti, E.P.; Reinhardt, R.; Minor, T.; Saner, F.H.; Canbay, A.; Treckmann, J.W.; Sotiropoulos, G.C.; et al. Donor information based prediction of early allograft dysfunction and outcome in liver transplantation. *Liver Int.* **2015**, *35*, 156–163. [CrossRef]
7. Minor, T.; Saad, S.; Nagelschmidt, M.; Koetting, M.; Fu, Z.; Paul, A.; Isselhard, W. Successful transplantation of porcine livers after warm ischemic insult in situ and cold preservation including postconditioning with gaseous oxygen. *Transplantation* **1998**, *65*, 1262–1264. [CrossRef]
8. Minor, T.; Paul, A. Hypothermic reconditioning in organ transplantation. *Curr. Opin. Organ Transplant.* **2013**, *18*, 161–167. [CrossRef]
9. Fuller, B.J.; Busza, A.L.; Proctor, E. Possible resuscitation of liver function by hypothermic reperfusion in vitro after prolonged (24-hour) cold preservation–a 31p nmr study. *Transplantation* **1990**, *50*, 511–512. [CrossRef] [PubMed]
10. Koetting, M.; Lüer, B.; Efferz, P.; Paul, A.; Minor, T. Optimal time for hypothermic reconditioning of liver grafts by venous systemic oxygen persufflation (vsop) in a large animal model. *Transplantation* **2011**, *91*, 42–47. [CrossRef] [PubMed]
11. Minor, T.; Koetting, M.; Kaiser, G.; Efferz, P.; Lüer, B.; Paul, A. Hypothermic Reconditioning by Gaseous Oxygen Improves Survival After Liver Transplantation in the Pig. *Am. J. Transplant.* **2011**, *11*, 2627–2634. [CrossRef] [PubMed]
12. Minor, T.; Stegemann, J.; Hirner, A.; Koetting, M. Impaired autophagic clearance after cold preservation of fatty livers correlates with tissue necrosis upon reperfusion and is reversed by hypothermic reconditioning. *Liver Transplant.* **2009**, *15*, 798–805. [CrossRef]
13. Kim, J.S.; NItta, T.; Mohuczy, D.; O'Malley, K.A.; Moldawer, L.L.; Dunn, W.A., Jr.; Behrns, K.E. Impaired autophagy: A mechanism of mitochondrial dysfunction in anoxic rat hepatocytes. *Hepatology* **2008**, *47*, 1725–1736. [CrossRef] [PubMed]
14. Gustafsson, Å.B.; Gottlieb, R.A. Recycle or Die: The Role of Autophagy in Cardioprotection. *J. Mol. Cell. Cardiol.* **2008**, *44*, 654–661. [CrossRef]
15. Treckmann, J.; Minor, T.; Saad, S.; Özcelik, A.; Malagó, M.; Broelsch, C.E.; Paul, A. Retrograde oxygen persufflation preservation of human livers: A pilot study. *Liver Transplant.* **2008**, *14*, 358–364. [CrossRef]

16. Khorsandi, S.E.; Jitraruch, S.; Fairbanks, L.; Cotoi, C.; Jassem, W.; Vilca-Melendez, H.; Prachalias, A.; Dhawan, A.; Heaton, N.; Srinivasan, P. The effect of anterograde persufflation on energy charge and hepatocyte function in donation after cardiac death livers unsuitable for transplant. *Liver Transplant.* **2014**, *20*, 698–704. [CrossRef]

17. Minor, T. Vascular oxygen persufflation for preservation and reconditioning of marginal liver grafts. In *Organ Preservation and Reengineering*, 1st ed.; Uygun, K., Lee, C.E., Eds.; Artech House: Norwood, MA, USA, 2011; pp. 125–134.

18. Minor, T.; Pütter, C.; Gallinat, A.; Ose, C.; Kaiser, G.; Scherag, A.; Treckmann, J.; Paul, A. Oxygen persufflation as adjunct in liver preservation (OPAL): Study protocol for a randomized controlled trial. *Trials* **2011**, *12*, 234. [CrossRef] [PubMed]

19. Olthoff, K.M.; Kulik, L.; Samstein, B.; Kaminski, M.; Abecassis, M.; Emond, J.; Shaked, A.; Christie, J.D. Validation of a current definition of early allograft dysfunction in liver transplant recipients and analysis of risk factors. *Liver Transplant.* **2010**, *16*, 943–949. [CrossRef] [PubMed]

20. Braat, A.E.; Blok, J.J.; Putter, H.; Adam, R.; Burroughs, A.K.; Rahmel, A.O.; Porte, R.J.; Rogiers, X.; Ringers, J.; European Liver and Intestine Transplant Association (ELITA) and Eurotransplant Liver Intestine Advisory Committee (ELIAC). The Eurotransplant Donor Risk Index in Liver Transplantation: ET-DRI. *Am. J. Transplant.* **2012**, *12*, 2789–2796. [CrossRef]

21. Charlson, M.E.; Pompei, P.; Ales, K.L.; MacKenzie, C. A new method of classifying prognostic comorbidity in longitudinal studies: Development and validation. *J. Chronic Dis.* **1987**, *40*, 373–383. [CrossRef]

22. Isselhard, W.; Witte, J.; Denecke, H.; Berger, M.; Fischer, J.H.; Molzberger, H.; Freiberg, C.; Ammermann, D.; Brunke, M. Function and metabolism of canine kidneys after aerobic ischemia by retrograde persufflation with gaseous oxygen. *Exp. Med.* **1974**, *164*, 35–44. [CrossRef]

23. Ross, H.; Escott, M.L. Gaseous oxygen perfusion of the renal vessels as an adjunct in kidney preservation. *Transplantation* **1979**, *28*, 362–364. [CrossRef]

24. Rolles, K.; Foreman, J.; Pegg, D.E. A pilot clinical study of retrograde oxygen persufflation in renal preservation. *Transplantation* **1989**, *48*, 339–342.

25. Minor, T.; Isselhard, W. Venous oxygen insufflation to prevent reoxygenation injury after ischemia of a solid organ. *Transplantation* **1994**, *58*, 121–123. [PubMed]

26. Minor, T.; Klauke, H.; Vollmar, B.; Isselhard, W.; Menger, M.D. Biophysical aspects of liver aeration by vascular persufflation with gaseous oxygen. *Transplantation* **1997**, *63*, 1843–1846. [CrossRef] [PubMed]

27. Minor, T.; Isselhard, W.; Klauke, H. Reduction in nonparenchymal cell injury and vascular endothelial dysfunction after cold preservation of the liver by gaseous oxygen. *Transpl. Int.* **1996**, *9*, S425–S428. [CrossRef]

28. Minor, T.; Klauke, H.; Nagelschmidt, M.; Isselhard, W. Reduction of proteolysis by venous-systemic oxygen persufflation during rat liver preservation and improved functional outcome after transplantation. *Transplantation* **1997**, *63*, 365–368. [CrossRef]

29. Minor, T.; Akbar, S.; Tolba, R.; Dombrowski, F. Cold preservation of fatty liver grafts: prevention of functional and ultrastructural impairments by venous oxygen persufflation. *J. Hepatol.* **2000**, *32*, 105–111. [CrossRef]

30. Dutkowski, P.; Polak, W.G.; Muiesan, P.; Schlegel, A.; Verhoeven, C.J.; Scalera, I.; DeOliveira, M.L.; Kron, P.; Clavien, P.-A. First comparison of hypothermic oxygenated perfusion versus static cold storage of human donation after cardiac death liver transplants: An international-matched case analysis. *Ann. Surg.* **2015**, *262*, 764–771. [CrossRef]

31. Hoyer, D.P.; Mathe, Z.; Gallinat, A.; Canbay, A.C.; Treckmann, J.W.; Rauen, U.; Paul, A.; Minor, T. Controlled oxygenated rewarming of cold stored livers prior to transplantation: First clinical application of a new concept. *Transplantation* **2016**, *100*, 147–152. [CrossRef] [PubMed]

32. Watson, C.J.E.; Kosmoliaptsis, V.; Randle, L.V.; Russell, N.K.; Griffiths, W.J.H.; Davies, S.; Mergental, H.; Butler, A.J. Preimplant Normothermic Liver Perfusion of a Suboptimal Liver Donated After Circulatory Death. *Am. J. Transplant.* **2015**, *16*, 353–357. [CrossRef] [PubMed]

bioengineering

MDPI

Review

Towards Bioengineered Liver Stem Cell Transplantation Studies in a Preclinical Dog Model for Inherited Copper Toxicosis

Hedwig S. Kruitwagen, Hille Fieten and Louis C. Penning *

Department of Clinical Sciences of Companion Animals, Faculty of Veterinary Medicine, Utrecht University, 3584CM Utrecht, The Netherlands; H.S.Kruitwagen@uu.nl (H.S.K.); H.Fieten@uu.nl (H.F.)
* Correspondence: L.C.Penning@uu.nl

Received: 26 June 2019; Accepted: 17 September 2019; Published: 25 September 2019

Abstract: Wilson Disease is a rare autosomal recessive liver disorder in humans. Although its clinical presentation and age of onset are highly variable, hallmarks include signs of liver disease, neurological features and so-called Kayser-Fleischer rings in the eyes of the patient. Hepatic copper accumulation leads to liver disease and eventually to liver cirrhosis. Treatment options include life-long copper chelation therapy and/or decrease in copper intake. Eventually liver transplantations are indicated. Although clinical outcome of liver transplantations is favorable, the lack of suitable donor livers hampers large numbers of transplantations. As an alternative, cell therapies with hepatocytes or liver stem cells are currently under investigation. Stem cell biology in relation to pets is in its infancy. Due to the specific population structure of dogs, canine copper toxicosis is frequently encountered in various dog breeds. Since the histology and clinical presentation resemble Wilson Disease, we combined genetics, gene-editing, and matrices-based stem cell cultures to develop a translational preclinical transplantation model for inherited copper toxicosis in dogs. Here we describe the roadmap followed, starting from the discovery of a causative copper toxicosis mutation in a specific dog breed and culminating in transplantation of genetically-engineered autologous liver stem cells.

Keywords: copper toxicosis; stem cell transplantation; Wilson Disease; preclinical large animal model

---

1. Introduction

The trace element copper is indispensable for various biochemical processes [1]. At the same time, the transition element copper (reduced as Cu^+ and oxidized as Cu^{2+}) is involved in chemical reactions leading to the production of reactive oxygen species. Therefore, its intracellular free concentrations need to be regulated within very narrow boundaries [2]. Regulation occurs at the level of cellular uptake, intracellular binding and distribution, and lastly cellular excretion. Copper is imported into cells mainly via Copper Transporter 1 (Ctr1) [3]. Once copper is inside the cell, copper binding proteins ensure that the free copper levels remain very low. These chaperone proteins include Cytochrome c Oxidase Copper Chaperone (Cox17), Copper Chaperone for Superoxide Dismutase (CCS), and Antioxidant protein1 (ATOX1) [1,2]. Intracellularly copper can be sequestered by glutathione, and metallothionein. Excretion is mediated via P-type ATPases, ATP7A and ATP7B [4]. Transport through the blood stream is mediated via ceruloplasmin. Copper related diseases in humans include Menke's Disease (copper deficiency disorder), Wilson Disease (copper accumulation), Indian childhood cirrhosis [5], endemic Tyrolean infantile cirrhosis [6], and idiopathic copper toxicosis [7]. The causative mutations for Wilson Disease are in the *ATP7B* gene [8,9]. The ATP7B protein is responsible for excretion of bound intrahepatic copper into the bile. Biliary copper excretion accounts for as much as 95% of the total body copper excretion. There is a wide variation in clinical presentation of Wilson Disease (WD) and a large number of mutations has been reported (well over 500; http://www.wilsondisease.med.ualberta.ca/database.asp) [10–12].

Phenotypical variation also occurs within the same genotype, hinting to the effect of modifier genes. In contrast to hepatic copper overload, Menkes Disease (MD) is presented with impaired copper absorption in various organs. Wilson Disease is a rare X-linked copper deficiency disorder caused by mutations in the *ATP7A* gene [13–17]. The limited genotype-phenotype correlation and the rarity of both WD and MD urge for innovative clinical approaches to improve the quality of life of people suffering from Wilson's or Wilson Disease. What makes dogs (*Canis lupus familiaris*), especially for WD, so well-suited? In order to fully appreciate the potential of these animals, some insights into the canine population structure are necessary. Ever since dogs were domesticated, they have been under severe artificial breeding selection, for instance for behavioral traits and/or specific morphological features [18]. This resulted in isolated genetic populations of dog breeds [19]. The limited genetic variation within breeds, and at the same time a large genetic variation over all breeds, provides a gold-mine for geneticists. Whereas the genetic variation over the various breeds remained intact, the reduced genetic variability within breeds worked as a genetic amplifier and offered researchers a genetic dissection microscope [19]. Together with the selection for a unique trait, such as excessive muscle formation, short limbs or a specific coat color, an increased risk for the development of specific disorders with a simple and/or complex inheritance pattern arose within breeds. Exploiting the downside of inbreeding may therefore be instrumental for the discovery of causative and modifier genes involved in complex diseases and/or rare diseases such as inherited copper toxicosis.

Next to the above mentioned genetic-argument, other research advantages reside within dogs. Dogs are of comparable size of humans and they share similar environmental exposures. Especially the size allows to design and test procedures at a humanized size with highly comparable anatomical arrangements. This is an obvious advantage for preclinical studies, for instance related to liver transplantation. In this respect, the readers might be aware that the first liver transplantations were performed in dogs [20].

In summary both genetic and technical arguments are in favor to utilize dogs as important preclinical models for inherited copper toxicosis. However, there is more to come. Veterinarians have been confronted with sheep and dogs presenting with copper related disorders already for decades [21–24]. Deleterious levels of hepatic copper are described in several dog breeds including Bedlington terriers, Skye terriers, West-Highland White terriers, Dobermanns, Dalmatians and Labrador retrievers [25–30]. Pedigree analysis of most breeds revealed a complex mode of inheritance of copper-mediated hepatitis. Therefore, the phenotypic expression is not dependent on one single genetic factor, but on mutations in more genes and also environmental factors are deemed influential in the phenotypic presentation. As an exception to the rule of thumb that copper toxicosis is a complex genetic disorder, a simple autosomal recessive mode of inheritance is observed in the Bedlington terrier [31].

2. A Roadmap towards a Relevant Preclinical Model Animal for Liver Stem Cell Transplantations

Although rodent models for WD have been instrumental to dissect molecularly how mutations in the *ATP7B* gene lead to hepatic copper accumulation, their size does not allow for longitudinal studies (individual animals followed for long time and consecutive measurements). This prompted us to investigate the genetic background of inherited copper toxicosis in several dog breeds. One important feature of a clinical model is the knowledge of the genetic cause and preferentially a simple breeding strategy to acquire sufficient number of experimental animals. In line with this, a similar progression of the disease in time strengthens the validity of the model. Another aspect is the feasibility to obtain sufficient and genetically gene-corrected liver stem cells, preferentially autologous to minimize the risk of rejection of the transplanted cells. Lastly, in view of the clinical application in human medicine, mode of cell transplantation must be similar to the one preferred in human medicine. All these steps will be outlined in more detail below. Figure 1 depicts the strategy followed for functional liver recovery after autologous genetically-engineered liver stem cell transplantation in a COMMD1 deficient dog.

A

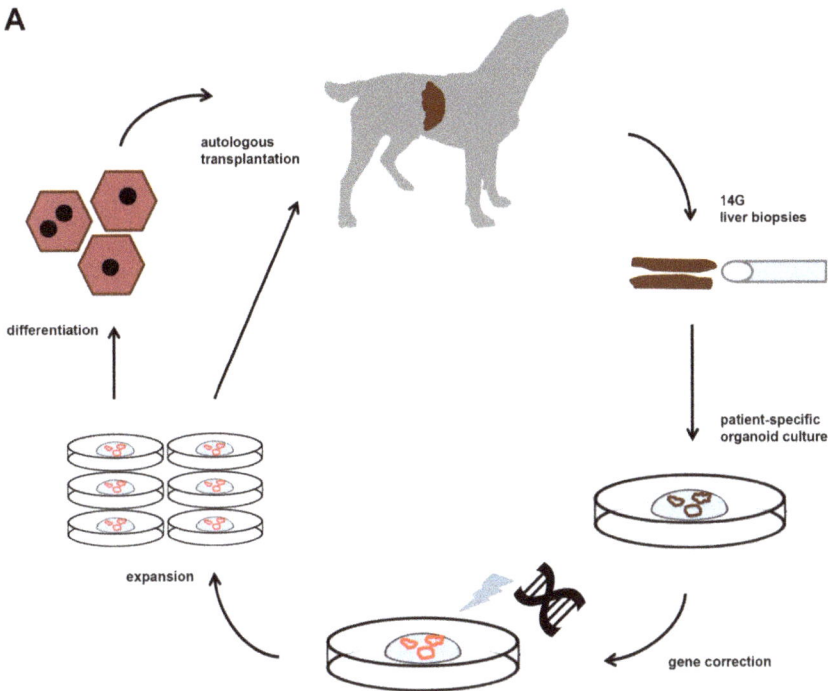

Figure 1. The strategy followed for functional liver recovery after autologous genetically-engineered liver stem cell transplantation in a COMMD1 deficient dog.

2.1. Requirement 1. Copper Accumulation in Bedlington Terriers Is caused by a Deletion of exon-2 in the COMMD1-gene

By means of mapping studies and positional cloning, a 13kB deletion covering exon-2 of the *commd1* gene was discovered as the causative mutation of Bedlington terrier copper toxicosis [31,32].

In the following years, the involvement of this mutation in other copper storage diseases was investigated. It turned out that this mutation was not causative for either Indian Childhood Cirrhosis (ICC), Endemic Tyrolean Infantile Cirrhosis (ETIC), nor Idiopathic copper toxicosis (ICT) [33].

Furthermore, whether or not the *murr1* mutations are somehow involved in WD is a matter of debate [34–36]. The intracellular interaction between the ATP7B protein and the COMMD1 protein (previously known as murr1) explains the similarities in WD and Bedlington terrier copper toxicosis [37]. The COMMD1 protein, COpper Metabolism Murr1 domain-containing protein had an unknown function at the time it was discovered. To unravel its function, among others yeast-two hybrid screens were used with COMMD1 as bait. Of interest, a direct COMMD1-ATP7B interaction occurs which was confirmed in cell lines [37]. In WD the interaction between ATP7B and COMMD1 is enhanced and leads to lower ATP7B stability. This interaction partially explains the similar phenotypes of WD in men and copper toxicosis in Bedlington terriers. At present a plethora of functions are related to the COMMD1 protein, including sodium transport via epithelial sodium channel (ENaC), trafficking of cystic fibrosis transmembrane conductance regulator (CFTR), inhibition of Cu/Zn -SOD, NFk-B signalling, Hypoxia Inducing Factor (HIF1) regulation and HIV-replication [37–41]. COMMD1 depletion leads to increased serum Low Density Lipoprotein (LDL) levels, due to mis localization of the LDL-receptor and consequently a reduced uptake of LDL particles [42]. One of the common themes of COMMD1 action seems to be related to protein degradation via ubiquitination, at least regarding NFkB-signalling, ENaC trafficking and HIF1alpha regulation. Results of immunoprecipitated ubiquitinated proteins with associated proteins have not been described, to our knowledge.

Two papers provided direct evidence for a crucial role in cellular copper regulation in in vitro cell cultures [41,43]. By means of siRNAs commd1 was silenced in HEK-293 (human embryonic kidney) cells and in BDE-cells, a canine liver cell line [41,43]. This resulted even in short term cell cultures for increased intracellular copper levels. The fetal lethality of COMMD1 -/- mice is most likely caused by defect in the placenta development, an organ with very high COMMD1 protein expression [43]. Finally, liver specific ATP7B deficient mice had increased hepatic copper levels [44], albeit not as high as the hepatic copper levels in Bedlington copper toxicosis (see Table 1). With respect to hepatic copper accumulation it turned out that COMMD1 protein functions as a chaperone not only for ATP7B, but also for ATP7A [45–47].

Table 1. Comparison of some parameters for Wilson Disease (WD) patients with potential large animal models such as Bedlington terriers (BT) and Labrador retrievers (LR).

	WD	**BT**	**LR**
Gene	*ATP7B*	*COMMD1*	*ATP7B/ATP7A*
Mode of inheritance	autosomal recessive	autosomal recessive	complex
Age of onset	variable	adolescence-mid age	adolescence-mid age
Liver pathology	cirrhosis	cirrhosis	cirrhosis
Hepatic Cu (mg/dwl)	<1000	<12,000	<1000
Neurology	impaired	not reported	not reported
Population	rare	rare *	very frequent
Kayser–Fleischer rings	present in 50%	not reported	not reported

mg/dwl means mg copper per kg dry weight liver; * due to negative breeding selection on copper toxicosis, the disease in Bedlington terriers almost disappeared.

In summary, the genetic cause of Bedlington terrier copper toxicosis is known, and despite not being in the same gene as for WD, protein–protein interaction of the WD-gene product and the Bedlington-gene product easily explain the similarities in hepatic copper accumulation.

2.2. Requirement 2. Longitudinal Studies on COMMD1 Deficient Dogs Highlight Similarities between WD and Canine Copper Toxicosis

All these in vitro and mouse models stimulated us perform longitudinal studies to describe in great detail how copper toxicosis progresses and to which extent this resembles WD. Therefore, an in-house breeding colony of five COMMD1 deficient dogs was followed for over four years. The simple mode of inheritance facilitated us to create a homozygous COMMD1 deficient breed on a Beagle background.

Biannual liver biopsies were taken for histology, copper measurements, immunohistochemistry, quantitative RT-PCR and Western blotting [48–50]. Although these animals are not geno-copies of human WD patients the disease progression at molecular and histological level clearly resembles WD. Variations between WD in men and COMMD1-deficiency mediated copper toxicosis in Bedlington terriers include the amount of copper accumulated (see Table 1) and the absence of neurological features and Kayser–Fleischer rings. Maximum copper accumulation was reached at 12 months of age (adolescence-mid age), which coincided with the first histological signs of hepatitis. At the same time, increased levels of *mt1A* (copper scavenger metallothionein) mRNA were observed. Slightly later, hepatic stellate cells became activated (α-SMA positivity), with increasing reticulin deposition and hepatocytic proliferation in later stages. A further increase over time of histologically confirmed hepatitis and pro-apoptotic caspase-3 activity (first noticed at 18 months) was observed. For further details on the temporal expression of genes involved in copper homeostasis and antioxidant mechanisms like *atox1* (antioxidant 1 copper chaperone), *ccs* (copper chaperone for cytochrome C oxidase), *cox17* (cyclooxygenase 17), *atp7A, atp7B, sod, cat*, and *gpx1* readers are referred to previous papers [49,50]. These longitudinal studies clearly established that COMMD1-deficient dogs develop copper-induced chronic liver disease and even cirrhosis in a comparable fashion as do human WD patients. Two minor variations are related to the age of onset which was much more standardized in the dogs and the copper accumulation was more extreme. Together, the important clinical and histological similarities positioned this breeding colony as genetically-defined large animal model to test clinical applicability of new therapeutics developed in rodent models.

2.3. Requirement 3. Culture of Sufficient Quantities of COMMD1-Functional Autologous Liver Stem Cells

The liver is one of the few organs acknowledged for its reparative capacity. Replication of differentiated hepatocytes and/or cholangiocytes are responsible for the volume regeneration to compensate for partially removed liver lobes. For this reason, the existence of liver stem cells seemed unnecessary. Some evidence even points to hepatocytes themselves as main sources of liver regeneration [51,52]. To complicate matters even more, several sources of hepatic stem cells are described, presumably depending on the model to induce hepatic damage or their involvement in the hepatocyte renewal during liver organ homeostasis [51–53]. Often one of the most obvious histological reactions is a proliferation of a subset of biliary cells, the so-called ductular reaction. Proposed stem cells include, among others, hepatic stellate cells, hepatocytes themselves, or self-renewing pericentral Axin2+ cells that differentiate into polyploid hepatocytes able to replace hepatocyte during homeostatic liver renewal [53–55]. Which of these various liver stem cells are involved in the daily wear-and tear of hepatocytes and cholangiocytes and which of these are hampered in their activation in case of severe liver damage is a matter of a fierce debate; the various points of view are summarized [56,57].

The presence and activation of liver progenitor cells was based on experimental mouse models, and ductular reactions (proliferation of intrahepatic bile ducts) were described in various diseases in humans [58–63]. Our initial papers described the activation of liver stem cells in canine hepatopathies and compared these patterns to the observation in human hepatopathies [64–67].

Having established that a ductular reaction occurred in diseased dog livers in a similar fashion as for human liver diseases, we designed experiments to culture canine liver progenitor cells.

Enrichment of liver progenitor cells by means of fluorescent activated cell sorting (FACS) revealed a side population with a gene expression pattern resembling progenitor cells [68].

Interestingly, during the culture of non-purified liver canine liver progenitor cells, on top of non-proliferating hepatocytes colonies of small, progenitor-like cells became visible [69]. Even more interesting these cells differentiated into a hepatocyte-like phenotype (e.g., albumin and MRP2 expression). This stimulated us to dissect pathways involved in their replication in order to obtain sufficient quantities of liver stem cells. Using a siRNA-screen targeting kinases, we established DYRK1A as a novel pathway specific for liver stem cell proliferation [70]. However, interference with harmine, a known inhibitor of DYRK1A, revealed little effect on cell proliferation as it turned out to be crucial for S-phase entry.

A recent development has been the establishment of liver stem cells cultures as 3D cultured organoids. Organoids are an artificially grown mass of (stem) cells resembling an organ's function. These stem cell based mini-organs are described from various organs including murine and human livers [71,72]. The clinical application of these mini-organs range from diseases modelling, advanced toxicology and pharmacological studies, and application in stem cell transplantations. For liver, dog, rat, and cat liver organoids have been described [73–75]. With these 3D in vitro cultures, expandable to almost infinity and at the same genetically stable, the combination of bioengineering and cell-biology becomes a reality. Indeed in 2015, canine liver organoids of COMMD1-deficient dogs were described in which by means of lentiviral transduction a COMMD1 cDNA was inserted, COMMD1 protein expression resumed in these organoids, as did their copper excretion and they survived under high copper culture conditions [73].

In brief, culture of large quantities of autologous liver stem cells with a functional copy of the COMMD1 cDNA was within reach.

2.4. Requirement 4. Number and Routing of Genetically-Engineered Autologous Transplanted Donor Cells

Hepatocyte transplantation in dogs was first described in 1996 preceding the clinical application of hepatocyte transplantations in humans [76]. Together with the few studies describing transplantation of healthy liver cells in human Crigler-Najjar syndrome, urea cycle defects and phenylketonuria we could make an estimated guess on the number of cells minimally required for functional liver recovery [77–80]. A growth stimulus for the transplanted cells was provided by a lobectomy of the recipient liver [80]. In order to avoid increased portal pressure during cell transplantation and to enhance engraftment, the required number of cells was injected via the portal vein on three consecutive days through an implanted port-a-cath system [81]. The portal vein has been the preferred routing in human patients to transplant liver cells in order to correct inborn errors of metabolism. This highlights another advantage of a large animal models over murine models, since portal vein injections in mice are very challenging.

3. Transplantation of Autologous COMMD1-Positive Liver Organoids into Copper-Laden Livers of COMMD1 Deficient Dogs

Having fulfilled relevant criteria for a valid preclinical model for stem cell transplantation we sailed out to culture enormous quantities of gene-corrected liver stem cells. In about 12 weeks of culture the number of gene-corrected liver stem cells was obtained. In this study, we are following five individual dogs up to two years after the transplantation, and the dogs' size permitted longitudinal liver sampling. All dogs tolerated the lobectomy and the subsequent liver cell transplantations.

Apparently the usage of the port-a-cath system did not reveal severe side-effects. At histology no signs of tumor formation caused by the transplanted cell are observed for the post-transplantation time points analyzed thus far.

4. Bedlington Terriers with a *commd1* Mutation or Labrador Retrievers with an *ATPB* Mutation, Which Is the Preferred Breed to Study WD?

Biomedical researchers are well aware of the fact that each model has its limitations, for instance it is a simplification or exaggeration of the reality. In other words, the beauty (and relevance) of a model is in the eye of the beholder. It is of utmost importance to make a rational decision on which large animal or which specific breed to be used for various preclinical studies to truly have impact on the quality of life of WD patients. In order to facilitate this decision, for instance if one would like to investigate the effect of novel therapeutic compounds, some aspects of two dog breeds with copper toxicosis are compared with clinical parameters of WD, as summarized in Table 1.

5. Future of Novel Preclinical Models, DoGtor Can You Help Me?

The European Commission recently established a reference network for rare liver diseases (ERN-RARE-LIVER). This shows Europe's perseverance to address the specific issues inherent to rare diseases, including limited research resources, a lack of scientific understanding and importantly a lack of public awareness. In this respect the participation of the Faculty of Veterinary Medicine (Utrecht, the Netherlands) in an EASL-sponsored consortium entitled Regenerative Hepatology is a crucial first step to bridge the scientific and preclinical gap for inherited copper toxicosis. Recently a recovery from acute liver failure without transplantation was reported in a small population ($n = 5$) [82]. Zinc and/or copper chelation contributed to the recovery. Similarly, a high zinc, low copper diet decreased hepatic copper levels in a subset of Labrador retrievers suffering from inherited copper toxicosis [83]. This shows the potential of comparative clinical studies in humans and dogs. However, it must be kept in mind that these animals are both target animal for therapy (pets as patients) and model animals. As for all models they represent a part of the complete picture, one or a few aspects are highlighted. In the case of hepatic stem cell transplantation, this is unlikely to become the main treatment of choice for WD since it will only affect liver function and not likely the neurological aspects.

Bioengineering **2019**, *6*, 88

Often research is of mice and meds, but we should consider for preclinical studies on bioengineered livers the option pets and vets. In other words, a dogmatic shift: bioengineered liver with large animal models will benefit people suffering from rare diseases, such as inherited copper toxicosis.

Author Contributions: Writing, H.S.K., H.F., and L.C.P.; Quantitative copper measurement, H.F.; transplantation studies H.S.K., and L.C.P.; Funding acquisition, L.C.P.

Funding: This research was funded by the Dutch Research Council NWO ZON/MW (116004121).

Conflicts of Interest: The authors declare no conflict of interest.

References

1. Inesi, G. Molecular features of copper binding proteins involved in copper homeostasis. Critical review. sometime be replace by from pipets-2-pets. In remains to be seen how the rational combination of IUBMB Life. *IUBMB Life* **2017**, *69*, 211–217.
2. Kim, B.E.; Nevitt, T.; Thiele, D.J. Mechanisms for copper acquisition, distribution and regulation. *Nat. Chem. Biol.* **2008**, *4*, 176–185. [PubMed]
3. Zhou, B.; Gitschier, J. hCTR1: A human gene for copper uptake identified by complementation in yeast. *Proc. Natl. Acad. Sci. USA* **1997**, *94*, 7481–7486. [CrossRef] [PubMed]
4. Lutsenko, S.; Barnes, N.L.; Bartee, M.Y.; Dmitriev, O.Y. Function and regulation of human copper-transporting ATPases. *Physiol. Rev.* **2007**, *87*, 1011–1046.
5. Tanner, M.S. Role of copper in Indian Childhood Cirrhosis. *Am. J. Clin. Nutr.* **1998**, *67*, S1074–S1081. [CrossRef] [PubMed]
6. Muller, T.; Feichtinger, H.; Berger, H.; Muller, W. Endemic Tyrolean Infantile Cirrhosis: An ecogenetic disorder. *Lancet* **1996**, *347*, 877–880. [PubMed]
7. Scheinberg, I.C.; Sternlieb, I. Wilson Disease and Idiopathic Copper Toxicosis. *Am. J. Cin. Nutr.* **1996**, *63*, S842–S845.
8. Bull, P.C.; Thomas, G.R.; Rommens, J.M.; Forbers, J.R.; Cox, D.W. The Wilson disease gene is a putative copper transporting P-type ATPase similar to the Menkes gene. *Nat. Genet.* **1993**, *5*, 327–337. [CrossRef]
9. Tanzi, R.E.; Petrukhin, K.; Chernov, I.; Pelleguer, J.L.; Wasco, W.; Ross, B.; Romano, D.M.; Parano, E.; Pavone, L.; Bzustowicz, L.M.; et al. The Wilson disease gene is a copper transporting ATPase with homology t the Menkes disease gene. *Nat. Genet.* **1993**, *5*, 344–350. [CrossRef]
10. Ferenci, P. Phenotype-genotype correlations in patients with Wilson's disease. *Ann. N. Y. Acad. Sci.* **2014**, *1315*, 1–5.
11. Lutsenko, S. Modifying factors and phenotypic diversity in Wilson's disease. *Ann. N. Y. Acad. Sci.* **2014**, *1315*, 56–63. [CrossRef] [PubMed]
12. Medici, V.; Weiss, K.H. Genetic and environmental modifiers of Wilson disease. In *Handbook of Clinical Neurology*; Elsevier: Amsterdam, the Netherlands, 2017; Volume 142, pp. 35–41.
13. Chelly, J.; Tumer, Z.; Tonnesen, T.; Petteson, A.; Ishikawa-Brush, Y.; Tommerup, N.; Norn, N.; Monaco, A.P. Isolation of a candidate gene for Menkes disease that encodes a potential heavy metal binding protein. *Nat. Genet.* **1993**, *3*, 14–19. [PubMed]
14. Mercer, J.F.B.; Livingston, J.; Hall, B.; Paynter, J.A.; Begy, C.; Chandrasekharappa, S.; Lockhart, P.; Grimes, A.; Bhave, M.; Siemieniak, D.; et al. Isolation of a partial candidate gene for Menkes disease by positional cloning. *Nat. Genet.* **1993**, *3*, 20–25. [PubMed]
15. Vulpe, C.; Levinson, B.; Whitney, S.; Packman, S.; Gitschier, J. Isolation of a candidate gene for Menkes disease and evidence that it encodes a copper-transporting ATPase. *Nat. Genet.* **1993**, *3*, 7–13. [PubMed]
16. Moller, L.B.; Mogensen, M.; Horn, N. Molecular diagnosis of Menkes disease: Genotype-phenotype correlation. *Biochimie* **2009**, *91*, 1273–1277. [CrossRef] [PubMed]
17. Tumer, Z. An overview and update of ATP7A mutations leading to Menkes disease and occipital horn syndrome. *Hum. Mutat.* **2013**, *34*, 417–429. [PubMed]
18. Larson, G.; Karlsson, E.K.; Perri, A.; Webster, M.T.; Ho, S.Y.; Peters, J.; Stahl, P.W.; Lingaas, F.; Fredholm, M.; Comstock, K.E.; et al. Rethinking dog domestication by integrating genetics, archeology, and biogeography. *Proc. Natl. Acad. Sci. USA* **2012**, *109*, 8878–8883. [CrossRef]

19. Parker, H.G.; Shearin, A.L.; Ostrander, E.A. Man's best friend becomes biology's best in show: Genome analyses in the domestic dog. *Annu. Rev. Genet.* **2010**, *44*, 309–336.

20. Starzl, T.E.; Kaupp, H.A., Jr.; Brock, D.R.; Linman, J.W. Studies on the rejection of the transplanted homologous dog liver. *Surg. Gynecol. Obstet.* **1961**, *112*, 135–144.

21. Haywood, S.; Muller, T.; Muller, W.; Heinz-Erian, P.; Tanner, M.S.; Ross, G. Copper-associated liver disease in north ronaldsay sheep: A possible animal model for non-wilsonian hepatic copper toxicosis of infancy and childhood. *J. Pathol.* **2001**, *195*, 264–269. [CrossRef]

22. Fuentalba, I.C.; Aburto, E.M. Animal models of copper-associated liver disease. *Comp. Hepatol.* **2003**, *2*, 5. [CrossRef]

23. Fieten, H.; Penning, L.C.; Leegwater, P.A.; Rothuizen, J. New canine models of copper toxicosis: Diagnosis, treatment, and genetics. *Ann. N. Y. Acad. Sci.* **2014**, *1314*, 42–48. [CrossRef] [PubMed]

24. Reed, E.; Lutsenko, S.; Bandmann, O. Animal models of Wilson disease. *J. Neurochem.* **2018**, *146*, 356–373. [CrossRef] [PubMed]

25. Twedt, D.C.; Sternlieb, I.; Gilbertson, S.R. Clinical, morphologic, and chemical studies on copper toxicosis of Bedlington Terriers. *J. Am. Vet. Med. Assoc.* **1979**, *175*, 269–275. [PubMed]

26. Haywood, S.; Rutgers, H.C.; Christian, M.K. Hepatitis and copper accumulation in Skye terriers. *Vet. Pathol.* **1988**, *25*, 408–414. [CrossRef] [PubMed]

27. Thornburg, L.P.; Rottinghaus, G.; Dennis, G.; Crawford, S. The relationship between hepatic copper content and morphologic changes in the liver of West Highland White Terriers. *Vet. Pathol.* **1996**, *33*, 656–661. [CrossRef]

28. Thornburg, L.P. Histomorphological and immunohistochemical studies of chronic active hepatitis in Doberman Pinschers. *Vet. Pathol.* **1998**, *35*, 380–385. [CrossRef]

29. Webb, C.B.; Twedt, D.C.; Meyer, D.J. Copper-associated liver disease in Dalmatians: A review of 10 dogs (1998-2001). *J. Vet. Intern. Med.* **2002**, *16*, 665–668.

30. Hoffmann, G.; van den Ingh, T.S.; Bode, P.; Rothuizen, J. Copper-associated chronic hepatitis in Labrador Retrievers. *J. Vet. Intern. Med.* **2006**, *20*, 856–861. [CrossRef]

31. Van De Sluis, B.; Rothuizen, J.; Pearson, P.L.; van Oost, B.A.; Wijmenga, C. Identification of a new copper metabolism gene by positional cloning in a purebred dog population. *Hum. Mol. Genet.* **2002**, *11*, 165–173. [CrossRef]

32. Van de Sluis, B.J.; Breen, M.; Nanji, M.; van Wolferen, M.; de Jong, P.; Binns, M.M.; Pearson, P.L.; Kuipers, J.; Rothuizen, J.; Cox, D.W.; et al. Genetic mapping of the copper toxicosis locus in Bedlington terriers to dog chromosome 10, in a region syntenic to human chromosome region 2p13-p16. *Hum. Mol. Genet.* **1999**, *8*, 501–507. [CrossRef]

33. Müller, T.; van de Sluis, B.; Zhernakova, A.; van Binsbergen, E.; Janecke, A.R.; Bavdekar, A.; Pandit, A.; Weirich-Schwaiger, H.; Witt, H.; Ellemunter, H.; et al. The canine copper toxicosis gene MURR1 does not cause non-Wilsonian hepatic copper toxicosis. *J. Hepatol.* **2003**, *38*, 164–168. [PubMed]

34. Stuehler, B.; Reichert, J.; Stremmel, W.; Schaefer, M. Analysis of the human homologue of the canine copper toxicosis gene MURR1 in Wilson disease patients. *J. Mol. Med.* **2004**, *82*, 629–634. [CrossRef] [PubMed]

35. Lovicu, M.; Dessi, V.; Lepori, M.B.; Zappu, A.; Zancan, L.; Giacchino, R.; Marazzi, M.G.; Iorio, R.; Vegnente, A.; Vajro, P.; et al. The canine copper toxicosis gene MURR1 is not implicated in the pathogenesis of Wilson disease. *J. Gastroenterol.* **2006**, *41*, 582–586. [PubMed]

36. Wu, Z.Y.; Zhao, G.X.; Chen, W.J.; Wang, N.; Wan, B.; Lin, M.T.; Murong, S.X.; Yu, L. Mutation analysis of 218 Chinese patients with Wilson diseases revealed no correlation between the canine copper toxicosis gene MURR1 And Wilson disease. *J. Mol. Med.* **2006**, *84*, 438442. [CrossRef] [PubMed]

37. De Bie, P.; van de Sluis, B.; Klomp, L.; Wijmenga, C. The many faces of the copper metabolism protein MURR1/COMMD1. *J. Hered.* **2005**, *97*, 803–811. [CrossRef]

38. Maine, G.N.; Burstein, E. COMMD proteins: COMMing to the scene. *Cell. Mol. Life Sci.* **2007**, *64*, 1997–2005.

39. Fedoseienko, A.; Bartuzi, P.; van de Sluis, B. Functional understanding of the versatile protein copper metabolism MURR1 domain 1 (COMMD1) in copper homeostasis. *Ann. N. Y. Acad. Sci.* **2014**, *1314*, 6–14.

40. Riera-Romo, M. COMMD1: A Multifunctional Regulatory Protein. *J. Cell. Biochem.* **2018**, *119*, 34–51.

41. Ganesh, L.; Burstein, E.; Guha-Niyogi, A.; Louder, M.K.; MAscola, J.R.; Klomp, L.W.; Wijmenga, C.; Duckett, C.S.; Nabel, G.J. The gene product Murr1 restricts HIV-1 replication in resting CD4+ lymphocytes. *Nature* **2003**, *426*, 853–857.

42. Bartuzi, P.; Billadeau, D.D.; Favier, R.; Rong, S.; Dekker, D.; Fedoseienko, A.; Fieten, H.; Wijers, M.; Levels, J.H.; Huijkman, N.; et al. CCC- and WASH-mediated endosomal sorting of LDLR is required for normal clearance of circulating LDL. *Nat. Commun.* **2016**, *7*, 10961. [CrossRef] [PubMed]

43. Spee, B.; Arends, B.; van Wees, A.M.; Bode, P.; Penning, L.C.; Rothuizen, J. Functional consequences of RNA interference targeting COMMD1 in a canine hepatic cell line in relation to copper toxicosis. *Anim. Genet.* **2007**, *38*, 168–170. [PubMed]

44. Vonk, W.I.; Bartuzi, P.; de Bie, P.; Kloosterhuis, N.; Wichers, C.G.; Berger, R.; Haywood, S.; Klomp, L.W.; Wijmenga, C.; van de Sluis, B. Liver-specific Commd1 knockout mice are susceptible to hepatic copper accumulation. *PLoS ONE* **2011**, *6*, e29183.

45. De Bie, P.; van de Sluis, B.; Burstein, E.; van den Berghe, P.V.; Muller, P.; Berger, R.; Gitlin, J.D.; Wijmenga, C.; Klomp, L.W. Distinct Wilson's disease mutations in ATP7B are associated with enhanced binding to COMMD1 and reduced stability of ATP7B. *Gastroenterology* **2007**, *133*, 1316–1326. [CrossRef] [PubMed]

46. Weiss, K.H.; Lozoya, J.C.; Tuma, S.; Gotthardt, D.; Reichert, J.; Ehehalt, R.; Stremmel, W.; Fullekrug, J. Copper-induced translocation of the Wilson disease protein ATP7B independent of Murr1/COMMD1 and Rab7. *Am. J. Pathol.* **2008**, *173*, 1783–1794. [CrossRef] [PubMed]

47. Vonk, W.I.; de Bie, P.; Wichers, C.G.; van den Berghe, P.V.; van der Plaats, R.; Berger, R.; Wijmenga, C.; Klomp, L.W.; van de Sluis, B. The copper-transporting capacity of ATP7A mutants associated with Menkes disease is ameliorated by COMMD1 as a result of improved protein expression. *Cell. Mol. Life Sci.* **2012**, *69*, 149–163. [PubMed]

48. Favier, R.P.; Spee, B.; Penning, L.C.; Rothuizen, J. Copper-induced hepatitis: The COMMD1 deficient dog as a translational animal model for human chronic hepatitis. *Vet. Q.* **2011**, *31*, 49–60.

49. Favier, R.P.; Spee, B.; Schotanus, B.A.; van den Ingh, T.S.; Fieten, H.; Brinkhof, B.; Viebahn, C.S.; Penning, L.C.; Rothuizen, J. COMMD1-deficient dogs accumulate copper in hepatocytes and provide a good model, for chronic hepatitis and fibrosis. *PLoS ONE* **2012**, *7*, e42158. [CrossRef]

50. Favier, R.P.; Spee, B.; Fieten, H.; van den Ingh, T.S.; Schotanus, B.A.; Brinkhof, B.; Rothuizen, J.; Penning, L.C. Aberrant expression of copper associated genes after copper accumulation in COMMD1-deficient dogs. *J. Trace Elem. Med. Biol.* **2015**, *29*, 347–353. [CrossRef]

51. Schaub, J.R.; Malato, Y.; Gormond, C.; Willenbring, H. Evidence against a stem cell origin of new hepatocytes in a common mouse model of chronic liver injury. *Cell Rep.* **2014**, *8*, 933–939. [CrossRef]

52. Yanger, K.; Knigin, D.; Zong, Y.; Maggs, L.; Gu, G.; Akiyama, H.; Pikarsky, E.; Stanger, B.Z. Adults hepatocytes are generated by self-duplication rather than stem cell differentiation. *Cell Stem Cell* **2014**, *15*, 340–349.

53. Kordes, C.; Sawitza, I.; Goetze, S.; Herebian, D.; Haussinger, D. Hepatic stellate cells contribute to progenitor cells and liver regeneration. *J. Clin. Investig.* **2014**, *124*, 5503–5515. [CrossRef] [PubMed]

54. Tarlow, B.D.; Pelz, C.; Naugler, W.E.; Wakefield, L.; Wilson, E.M.; Finegold, M.J.; Grompe, M. Bipotent adult liver progenitors are derived from chronically injured mature hepatocytes. *Cell Stem Cell* **2014**, *15*, 605–618. [PubMed]

55. Wang, B.; Zhao, L.; Fish, M.; Logan, C.Y.; Nusse, R. Self-renewing diploid axin2+ cells fuel homeostatic renewal of the liver. *Nature* **2015**, *524*, 180–185. [PubMed]

56. Huch, M.; Dolle, L. The plastic cellular states of liver cells: Are EpCAM and Lgr5 fit for purpose? *Hepatology* **2016**, *64*, 652–662.

57. Huch, M. The versatile and plastic liver. *Nature* **2015**, *517*, 155–156. [PubMed]

58. De Vos, R.; Desmet, V. Ultrastructural characteristics of novel epithelial cell types identified in human pathologic liver specimens with chronic ductular reaction. *Am. J. Pathol.* **1992**, *140*, 1441–1450. [PubMed]

59. Hsia, C.C.; Evarts, R.P.; Nakatsukasa, H.; Marsden, E.R.; Thorgeirsson, S.S. Occurrence of oval-type cells in hepatitis B virus-associated human hepatocarcinogenesis. *Hepatology* **1992**, *16*, 1327–1333.

60. Roskams, T.; de Vos, R.; Desmet, V. "Undifferentiated progenitor cells" in focal nodular hyperplasia of the liver. *Histopathology* **1996**, *28*, 291–299. [CrossRef]

61. Crosby, H.A.; Hubscher, S.; Fabris, L.; Joplin, R.; Sell, S.; Kelly, D.; Strain, A.J. Immunolocalization of putative human liver progenitor cells in livers from patients with end-stage primary biliary cirrhosis and sclerosing cholangitis using the monoclonal antibody OV-6. *Am. J. Pathol.* **1998**, *152*, 771–779.

62. Lowes, K.N.; Brennan, B.A.; Yeoh, G.C.; Olynyk, J.K. Oval cell numbers in human chronic liver diseases are directly related to disease severity. *Am. J. Pathol.* **1999**, *154*, 537–541. [PubMed]

63. Libbrecht, L.; Desmet, V.; Van Damme, B.; Roskams, T. The immunohistochemical phenotype of dysplastic foci in human liver: Correlation with putative progenitor cells. *J. Hepatol.* **2000**, *33*, 76–84. [CrossRef]

64. Schotanus, B.A.; van den Ingh, T.S.; Penning, L.C.; Rothuizen, J.; Roskams, T.A.; Spee, B. Cross-species immunohistochemical investigation of the activation of the liver progenitor cell niche in different types of liver disease. *Liver Int.* **2009**, *29*, 1241–1252. [CrossRef] [PubMed]

65. Ijzer, J.; Schotanus, B.A.; Vander Borght, S.; Roskams, T.A.; Kisjes, R.; Penning, L.C.; Rothuizen, J.; van den Ingh, T.S. Characterisation of the hepatic progenitor cell compartment in normal liver and in hepatitis: An immunohistochemical comparison between dog and man. *Vet. J.* **2010**, *184*, 308–314. [PubMed]

66. Kruitwagen, H.S.; Spee, B.; Viebahn, C.S.; Venema, H.B.; Penning, L.C.; Grinwis, G.C.; Favier, R.P.; van den Ingh, T.S.; Rothuizen, J.; Schotanus, B.A. The canine hepatic progenitor cell niche: Molecular characterisation in health and disease. *Vet. J.* **2014**, *201*, 345–532. [CrossRef] [PubMed]

67. Kruitwagen, H.S.; Spee, B.; Fieten, H.; van Steenbeek, F.G.; Schotanus, B.A.; Penning, L.C. Translation from mice to men: Are dogs a dodgy intermediate? *Eur. Med. J. Hepatol.* **2014**, *1*, 48–54.

68. Arends, B.; Vankelecom, H.; Vander Borght, S.; Roskams, T.; Penning, L.C.; Rothuizen, J.; Spee, B. The dog liver contains a "side population" of cells with hepatic progenitor-like characteristics. *Stem Cells Dev.* **2009**, *18*, 343–350. [CrossRef] [PubMed]

69. Arends, B.; Spee, B.; Schotanus, B.A.; Roskams, T.; van den Ingh, T.S.; Penning, L.C.; Rothuizen, J. In vitro differentiation of liver progenitor cells derived from healthy dog livers. *Stem Cells Dev.* **2009**, *18*, 351–358. [PubMed]

70. Kruitwagen, H.S.; Westendorp, B.; Viebahn, C.S.; Post, K.; van Wolferen, M.E.; Oosterhoff, L.A.; Egan, D.A.; Delabar, J.M.; Toussaint, M.J.; Schotanus, B.A.; et al. DYRK1A is a regulator of S-phase entry in hepatic progenitor cells. *Stem Cell Dev.* **2018**, *27*, 133–146.

71. Huch, M.; Dorrell, C.; Boj, S.F.; van Es, J.H.; Li, V.S.; van de Wetering, M.; Sato, T.; Hamer, K.; Sasaki, N.; Finegold, M.J.; et al. In vitro expansion of single Lgr5+ liver stem cells induced by Wnt-driven regeneration. *Nature* **2013**, *494*, 247–250. [CrossRef]

72. Huch, M.; Gehart, H.; van Boxtel, R.; Hamer, K.; Blokzijl, F.; Verstegen, M.M.A.; Ellis, E.; van Wenum, M.; Fuchs, S.A.; de Ligt, J.; et al. Long-term culture of genome-stable bipotent stem cells from adult human liver. *Cell* **2015**, *160*, 299–312. [CrossRef] [PubMed]

73. Nantasanti, S.; Spee, B.; Kruitwagen, H.S.; Chen, C.; Geijsen, N.; Oosterhoff, L.A.; van Wolferen, M.E.; Palaez, N.; Fieten, H.; Wubbolts, R.W.; et al. Disease Modeling and Gene Therapy of Copper Storage Disease in Canine Hepatic Organoids. *Stem Cell Rep.* **2015**, *5*, 895–907. [CrossRef] [PubMed]

74. Kuijk, E.W.; Rasmussen, S.; Blokzijl, F.; Huch, M.; Gehart, H.; Toonen, P.; Begthel, H.; Clevers, H.; Geurts, A.M.; Cuppen, E. Generation and characterization of rat liver stem cell lines and their engraftment in a rat model of liver failure. *Sci. Rep.* **2016**, *6*, 22154. [CrossRef] [PubMed]

75. Kruitwagen, H.S.; Oosterhoff, L.A.; Vernooij, I.G.W.H.; Schrall, I.M.; van Wolferen, M.E.; Bannink, F.; Roesch, C.; van Uden, L.; Molenaar, M.R.; Helms, J.B.; et al. Long-term adult feline liver organoids cultures for disease modeling of hepatic steatosis. *Stem Cell Rep.* **2017**, *8*, 822–830.

76. Kocken, J.M.; Borel Rinkes, I.H.; Bijma, A.M.; de Roos, W.K.; Bouwman, E.; Terpstra, O.T.; Sinaasappel, M. Correction of an inborn error of metabolism by intraportal hepatocyte transplantation in a dog model. *Transplantation* **1996**, *62*, 358–364. [CrossRef]

77. Fox, I.J.; Chowdhury, J.R.; Kaufman, S.S.; Goertzen, T.C.; Chowdhury, N.R.; Warkentin, P.I.; Dorko, K.; Sauter, B.V.; Strom, S.C. Treatment of the Crigler-Najjar syndrome type I with hepatocyte transplantation. *N. Eng. J. Med.* **1998**, *338*, 1422–1426.

78. Puppi, J.; Tan, N.; Mitry, R.R.; Highes, R.D.; Lehec, S.; Miele-Vergani, G.; Karani, J.; Champion, M.P.; Heaton, N.; Mohamed, R.; et al. Hepatocyte transplantation followed by auxiliary liver transplantation; a novel treatment of ornithine transcarbamylase deficiency. *Am. J. Transplant.* **2008**, *8*, 452–457. [CrossRef]

79. Stephenne, X.; Najimi, M.; Sibille, C.; Nasogne, M.C.; Smets, F.; Sokal, E.M. Sustained engraftment and tissue enzyme activity after liver cell transplantation for arginosuccinate lyase deficiency. *Gastroenterology* **2006**, *130*, 137–1323. [CrossRef]

80. Stephenne, X.; Debray, F.G.; Smets, F.; Jazouli, N.; Sana, G.; Tondreau, T.; Menten, R. Hepatocyte transplantation using the domino concept in a child with tetrabiopterin non responsive phenylketonuria. *Cell Transplant.* **2012**, *21*, 2765–2770.

81. Guha, C.; Deb, N.J.; Sappal, B.S.; Ghosh, S.S.; Roy-Chowdhury, N.; Roy-Chowdhury, J. Amplification of engrafted hepatocytes by preparative manipulation of the hoist liver. *Artif. Organs* **2001**, *25*, 522–528. [CrossRef]

82. Darwish, A.A.; Sokal, E.; Stephenne, X.; Najimi, M.; de Goyet, J.D.V.; Reding, R. Permanent access to the portal system for cellular transplantation using an implantable port device. *Liver Transplant.* **2004**, *10*, 1213–1215. [CrossRef] [PubMed]

83. Fieten, H.; Biourge, V.C.; Watson, A.L.; Leegwater, P.A.; van den Ingh, T.S.; Rothuizen, J. Dietary management of Labrador retrievers with subclinical hepatic copper accumulation. *J. Vet. Intern. Med.* **2015**, *29*, 822–827. [CrossRef] [PubMed]

bioengineering

MDPI

Review

Hydrogels for Liver Tissue Engineering

Shicheng Ye, Jochem W.B. Boeter, Louis C. Penning, Bart Spee and Kerstin Schneeberger *

Department of Clinical Sciences of Companion Animals, Faculty of Veterinary Medicine, Utrecht University, 3584 CT Utrecht, The Netherlands
* Correspondence: k.schneeberger@uu.nl

Received: 1 June 2019; Accepted: 3 July 2019; Published: 5 July 2019

Abstract: Bioengineered livers are promising in vitro models for drug testing, toxicological studies, and as disease models, and might in the future be an alternative for donor organs to treat end-stage liver diseases. Liver tissue engineering (LTE) aims to construct liver models that are physiologically relevant. To make bioengineered livers, the two most important ingredients are hepatic cells and supportive materials such as hydrogels. In the past decades, dozens of hydrogels have been developed to act as supportive materials, and some have been used for in vitro models and formed functional liver constructs. However, currently none of the used hydrogels are suitable for in vivo transplantation. Here, the histology of the human liver and its relationship with LTE is introduced. After that, significant characteristics of hydrogels are described focusing on LTE. Then, both natural and synthetic materials utilized in hydrogels for LTE are reviewed individually. Finally, a conclusion is drawn on a comparison of the different hydrogels and their characteristics and ideal hydrogels are proposed to promote LTE.

Keywords: hydrogel; tissue engineering; liver; bioengineered organ

1. Introduction

Liver tissue engineering (LTE) aims to construct liver models that mimic the functions of an in vivo liver as closely as possible. LTE has two main applications: First, as in vitro models, bioengineered livers can be used for testing of xenobiotics (e.g., drugs and pathogens), toxicological studies and as (patient-specific) disease models [1]. Ethical and practical issues hamper to conduct research on drugs and pathogens with living human beings; on the other hand, in vitro models, either hepatoma cell lines or primary human hepatocytes, cannot represent the true in vivo characteristics, where liver cells are spatially localized and cell polarity provides dynamic cues for cellular activities [2,3]. Thus, LTE could be used for drug development and toxicity testing [4] and as cell models for pathogen testing. Second, although currently far from clinical application, LTE aims to develop alternatives to donor organs for in vivo transplantations. Liver diseases are a major concern as they account for millions of deaths annually and the incidence of liver disease is still increasing worldwide [5]. End-stage liver disease or liver failure is the direct cause of death and the only curative option is orthotopic liver transplantation (OLT) [6–8]. However, donor shortage has restricted this treatment severely and many patients die while on the waiting list for applicable donor livers [9,10]. To solve the problem of donor shortage, hopes are that bioengineered livers could be an alternative in the future, and LTE is an essential approach to fabricate bioengineered livers.

Cell sources and supportive materials are the most fundamental ingredients for LTE. First of all, hepatic cells are indispensable and there are already several possible cell sources [5,7,8]. Primary hepatocytes are typically selected as the cell source [11–15] but are limited by the availability of primary tissue, the difficulty in maintaining the hepatic phenotype, and expanding the cells sufficiently [16,17]. Therefore, stem cells or progenitor cells that differentiate into the hepatic lineage are a viable alternative [18–20], and methods to expand induced pluripotent stem cells (iPSCs) or adult stem cell-derived hepatic cells have also been established [21–23]. The maturation status and hence function of stem cell derived hepatic cells do not reach primary hepatocyte levels yet, but can presumably be increased in the future by a combination of several maturation approaches [24]. Additionally, several groups have recently developed techniques which now allow for efficient in vitro expansion of primary human hepatocytes [25–27]. Now that methods have been developed for long-term culture of cells with hepatocyte function, there is a clear need to optimize biomaterials aiming to assemble various liver cell types properly.

Hydrogels are one of the most promising candidates to serve as supportive biomaterials and have been frequently used in tissue engineering and regenerative medicine (TERM) [28]. There are ample reviews or articles describing a wide variety of hydrogels [29–31]. Most of them are only focused on specific biomaterials such as nanocellulose [32], fibrin [33], collagen [34], poly(e-caprolactone) (PCL) [35], and poly(vinyl alcohol) (PVA) [36], and discuss the design methods [37,38] or proposed possible applications in TERM [39,40]. However, there is no clear statement on the different hydrogels used for LTE. Even though great improvements have been achieved, there are still no hydrogels available that mimic liver extracellular matrix (ECM) functionally, restricting LTE for both in vitro models and in vivo transplantation. Here, we compare different hydrogels used in LTE, and suggest possible applications.

2. Liver and LTE

2.1. Liver Functions and LTE

The main goal of LTE is to recapitulate main liver functions, not necessarily the liver architecture per se. The liver originates from the endoderm in the embryonic foregut [41] and is the largest internal organ in the human body, accounting for 2–5% of the body weight. It performs a complex array of more than 500 functions, including metabolic, synthetic, immunologic, and detoxification processes [8]. The most essential activities of the liver are to maintain an active urea cycle, albumin synthesis and drug metabolism as well as regulating whole-body metabolism and xenobiotic detoxification [42]. The liver has to face challenges daily while performing those vital functions, which may result in diseases caused by toxins, drugs, and viruses [8,9]. In addition, autoimmune diseases and liver cancer occur frequently [8,43]. These diseases can impair liver function and eventually lead to end-stage liver disease. Luckily, the liver has tremendous capacity to regenerate [44]. In the past decades, a comprehensive understanding of the mechanisms of liver regeneration has been established and a dozen reviews [42,44–54] have shown many different aspects of liver regeneration. Nevertheless, in many clinical scenarios liver regeneration is not sufficient to circumvent loss of a large volume of hepatic tissue [55]. LTE can on the one hand provide in vitro models for a better understanding of the pathophysiology of such liver diseases, and thereby contribute to the development of new treatment options. On the other hand, LTE might provide a treatment by itself in the future, and many groups have started to investigate the possibility of LTE for the creation of suitable liver transplants.

2.2. Liver Histology and LTE

The liver is one of the most complex organs in the human body (Figure 1). The mature human liver is composed of four lobes and structurally and histologically, the liver can be divided into four tissue systems [56]: intrahepatic vascular system, stroma, sinusoidal cells, and hepatocytes. Those tissue systems are made from multiple cell types, including the parenchymal cells, hepatocytes, and cholangiocytes, together with various non-parenchymal cells [57,58]. Hepatocytes constitute ~80% of the liver mass. The remaining part is made up by non-parenchymal cells, including liver sinusoidal endothelial cells, Kupffer cells, lymphocytes and stellate cells [44,59]. Although they take up a small portion of the liver volume (6.5%), they constitute 30–40% of the total cell number [6]. Those cell types enable the liver to exhibit a hierarchical structure consisting of repeated functional tissue units, the liver lobules. Within a lobule, a smaller amount of oxygenated blood enters through branches of the hepatic artery and the largest amount of low oxygenated blood enters through the portal vein and flows in specialized sinusoidal vessels toward the central vein. Bile, which is produced and excreted by hepatocytes, flows in the opposite direction towards the intrahepatic bile duct. Hepatocytes are polarized epithelial cells that interact closely with a number of nonparenchymal cell types along the sinusoidal tracts of the liver lobule. Collectively, these cellular components and multiscale tissue structures contribute to the diverse functional roles of the liver [8].

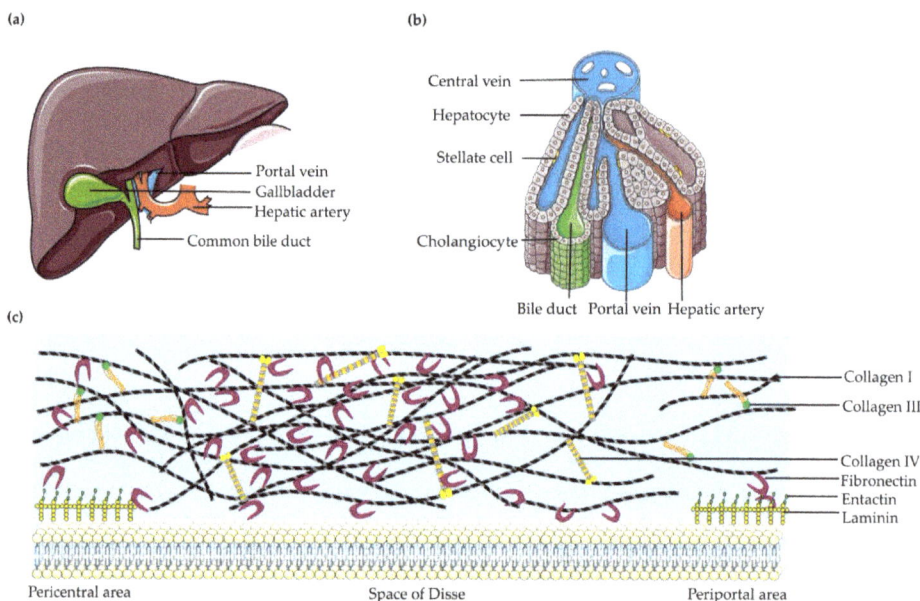

Figure 1. Liver histology and extracellular matrix (ECM). (**a**) A schematic representation of the whole human liver; (**b**) Schematic overview of the liver lobule; (**c**) The connection among major liver ECM components seen within the space of Disse.

Depending on the application of a bioengineered liver, it might not be necessary to recapitulate this entire complexity of the liver in LTE. For example, to study the pathophysiology of alpha1-antitrypsin (A1AT) deficiency, a purely epithelial liver model containing hepatocyte-like cells seems sufficient [23]. In general, however, a close-to-physiological 3D organization, cell composition and ECM has been shown to significantly improve the maturation and function of bioengineered tissues [24]. Most importantly, the cellular interactions [8] of the liver have to be established in order to create a structure that is similar to the native liver in both mechanism and function.

2.3. Liver ECM and LTE

Mimicking the liver ECM is another indispensable constitution for LTE. Although the ECM is only a small component of the liver, less than 3% of the relative area on a normal liver section [60], it has a crucial role [61]. The ECM provides cohesiveness within tissue compartments, induces polarization of cells, and acts as a major determinant of gene expression and differentiation [62,63]. As the major component of stroma [6], the liver ECM, mainly located at the interface between the blood flow and the epithelial compartment, plays a vital role in supporting and connecting hepatic cells, and also fulfills a big role in the polarity of parenchymal cells and thus the liver function. There are differences among ECM distributions of different areas in the adult liver. The liver can be divided into four major compartments: capsule, portal spaces, lobular interstitium (subsinusoidal space or space of Disse), and central space. The unique nature of the liver ECM is seen in the special configuration of the space of Disse. The liver lobule has no basement membrane (BM) and only an attenuated ECM consisting mostly of fibronectin, some collagen type I, and minor quantities of collagen types III, IV, V, and VI [61]. The structure and composition varies greatly in diseased livers [64–67]. Under normal conditions, the liver ECM consists of collagens type I and III (large fibrils), IV (net structure), V and VI (small fibrils), glycoproteins (laminin and fibronectin), elastins, glycosaminoglycans, and proteoglycans [68]. In fibrotic liver the ECM components are similar to those present in normal liver but are quantitatively increased (three- to five-fold increase in ECM) [64]. When liver damage is present, the liver ECM is produced mainly by hepatic stellate cells [6], the major fibrogenic cell type in human liver [69]. Even though fibrous tissue is quantitatively very limited in liver [64], the liver ECM forms the fibrous scaffold, provides a surface for cell adhesion, space for cell growth and migration, interacts with liver progenitor cells [70], and consists mostly of fibronectin, laminins, collagens, and signaling molecules [65,67]. Therefore, any modification in the liver ECM has a direct effect on liver structure and functions [64,71], which underlines the importance to mimic the liver ECM in LTE.

3. Hydrogels for LTE

Hydrogels are a promising candidate to mimic the liver ECM functionally in LTE. A hydrogel is a network of natural or synthetic hydrophilic polymer chains possessing a degree of flexibility similar to natural tissues. The term "hydrogel" first appeared in literature in 1894 [72] and the first generation of hydrogels were developed around 1960s, when poly(vinyl alcohol) (PVA) [73] and poly(2-hydroxyethl methacrylate) (pHEMA) [74] were described for the first time in publications. After the development for three generations, hydrogels are progressing to smart materials [75]. In order to find suitable hydrogels for LTE, various materials have been tested, but to date there is no hydrogel that mimics liver ECM adequately. Here, significant properties for hydrogels to mimic the liver ECM are introduced, together with hydrogels frequently utilized for LTE, in order to provide insights into hydrogels for LTE.

3.1. Properties Significant for LTE

To make the best use of different hydrogels, comprehensive understanding of their characteristics is necessary to mimic the liver ECM that is responsive for liver cell engraftment, long-term survival and function [76]. Those characteristics determine their various properties and several pivotal properties for LTE have been emphasized in the hope of optimizing the most suitable hydrogels. Properties of the ideal scaffold for LTE have been listed by Vasanthan et al. [77]. Here, those properties are integrated into two basic groups: biological properties [78] and physicochemical properties.

3.1.1. Biological Properties

The most fundamental characteristics of hydrogels for LTE are appropriate biological properties. Biological properties, such as biocompatibility [79,80], biodegradability, and bioactivity, have always received great attention when a hydrogel is used for TERM [81]. For example, cellular biocompatibility makes the nanofibrillar cellulose hydrogel suitable for the proliferation and differentiation of human hepatic cell lines [82]. Biodegradability makes hydrogels promising in applications on transplantation purposes [83]. Biodegradable hydrogels can not only act as the supportive scaffold for cells to perform many kinds of activities and form desirable tissue, they also provide the possibility to be cleared locally by enzymes that are specific to degrade those biomaterials [81]. The degradation speed can be regulated by the polymerization of the hydrogel.

Nevertheless, biocompatibility and biodegradability is not enough for hydrogels to support liver functions for LTE. They should also be bioactive, which means that the hydrogels are capable of transmitting dynamic signals instantly and are able to perform a variety of stimuli responses properly. To obtain these characteristics, spatiotemporal control of functional domains is needed so that the individual cell fate can be decided properly [84]. Thus, suitable hydrogels will act as bridges among cells as well as providing a "transportation system" within bioengineered tissues.

3.1.2. Physicochemical Properties

Physicochemical properties are as significant as biological properties to biomaterials [85]. It has been recognized that physical parameters are important determinants for cell growth and phenotype regulation [85]. For instance, Jeremy Bomo et al. demonstrated that the proliferation rates of normal and transformed hepatocytes are strongly induced by matrix with a higher stiffness [86]. Another study demonstrated that primary hepatocyte functions were preserved when cultured on matrix of normal liver stiffness (400–600 Pa) but significantly reduced when cultured on matrix with the stiffness of fibrotic liver (1.2–1.6 kPa) [87]. In order to form an efficient "transportation system", hydrogels have to gain more applicable mechanical properties besides suitable stiffness [88] such as mechanical stress and strength [79], elasticity and swelling, viscosity [83,89], and porosity [90,91]. For instance, hepatic cells are deposited within liver tissue with the stiffness around 640 Pa [92]. The pore size and porosity of scaffolds play an important role in the diffusion of growth factors and induce vascularization thereby aiding maintenance of liver specific functions [77]. As hepatocytes consume 5- to 10-fold more oxygen compared to other cells [93,94], pore size is a crucial factor which controls the mass transport of oxygen and nutrients into the interior of the scaffold, thereby supporting cellular growth in the region [95]. Porous scaffolds with pore sizes ranging from 50 to 150 μm and high inter-pore connectivity are desirable for the culture of hepatocytes [96]. Compared to hepatocytes cultured in the control scaffolds with non-uniform distribution of pores, primary hepatocytes cultured in a porous scaffold, owning a high porosity of around 83% with interconnected pores (average pore diameter 40–70 μm), showed an increase in albumin secretion and urea synthesis [97].

Apart from cellular and external influences, mechanical properties will also be affected by the materials themselves. In the in vivo ECM, the mechanical properties are largely influenced by proteoglycans and fibrous proteins. In the in vitro imitated ECM or implanted hydrogels, the mechanical properties are often influenced by the type and density of crosslinks. As the mechanics of the hydrogels affect the cell behavior and cell fate, mechanically patterned hydrogels have been created through local light exposure. Other influence factors include controllable variables such as concentration, polymer length and temperature [98]. On the other hand, natural ECMs have mechanical properties in a dynamic manner. Thus, hydrogel systems are designed with reversible mechanical properties to provide cells with optimal microenvironment in a spatiotemporal manner.

3.2. Categories of Hydrogels

Hydrogels could be distinguished with various parameters such as the preparation method, the overall charge, and the mechanical and structural characteristics. Here, the hydrogels are divided into two categories according to their origins: natural or synthetic.

3.2.1. Natural Hydrogels

Natural hydrogels originate from organisms and have natural advantages to mimic the ECM better when compared to synthetic hydrogels. Generally, natural hydrogels function well for common uses such as cell culture, drug delivery, and tissue engineering. Several natural hydrogels have been used for LTE, including collagen, gelatin, hyaluronan, fibrin, alginate, chitosan, polyhydroxyalkanoates, cellulose, and agarose. Here we specify which natural hydrogels have been used (Table 1) and which main advantages and/or disadvantages exist for LTE.

Table 1. Characteristics and applications of natural materials.

Natural Materials	Major Concerned Properties					Other Characteristics	Applications	References
	Biocompatible	Biodegradable	Hydrophilic	Thermal-Responsible	FDA			
HA	Yes	Yes, hyaluronidases	Yes	Yes	Approved	Nonimmunogenic, non-adhesive, good swelling and creep compliance properties, photopolymerizable, promote cell mortality and proliferation, reduces long-term inflammation, hepatic elimination	Tissue engineering, wound healing, angiogenesis, etc.	[99,100]
Alginate	Yes	Yes, controllable	Yes	Yes	Approved	Nonimmunogenic, bioactive, ease of gelation, suitable for in situ injection; poor elasticity, poor cell adhesion, mechanical weakness, difficulties in handling and sterilization	Tissue engineering and regeneration, as model ECMs, drug delivery	[101,102]
Collagen	Yes	Yes	Yes	Yes	Approved	Good permeability, low immunogenicity, poor mechanical properties	Tissue engineering (including cartilage, ligament, vessel etc.)	[34,103]
Gelatin	Yes	Yes, very fast	Yes	Yes	Approved	Ease of manipulation, high mechanical properties, easy to form films and matrix hydrogels, very viscous polymer, low thermal stability, low cost	Tissue engineering, drug discovery	[104,105]
Cellulose	Yes	Yes	Yes	Yes	Approved	Noncytotoxic, good thermal and mechanical properties, hydrogels with a high water content	Various derivatives in biomedical field	[32]
Matrigel	Yes	Yes	Yes	Yes	No	Bioactive, mechanical weakness, batch-to-batch variation, undefined composition	3D models, cell culture, mimic ECMs	[106–108]
Chitosan	Yes	Yes, lysozyme	Yes	Yes	Approved	Nonimmunogenic, good host response, high antimicrobial activity, very viscous polymer solution and pH-responsive, sufficient mechanical properties	Tissue engineering, e.g., liver, bone, skin, vessels	[31,89]
Agarose	Yes	Yes	Yes	Yes	No	High mechanical strength, ability to maintain the cellular phenotype	Mimics the microenvironment for hepatocytes	[97]
Fibrin	Yes	Yes	No	Yes	Approved	Easily autologous isolation, uniform cell distribution, limited mechanical strength, inflammatory response	Tissue engineering scaffolds, blood blotting, fertility preservation	[33,109,110]
PHAs	Yes	Yes	No	Yes	No	Non-toxic, piezoelectric properties, brittleness, tendency to crystallize	Tissue engineering including LTE, drug carrier, would healing	[29,111]

Abbreviations: Hyaluronic acid, HA; polyhydroxyalkanoates, PHAs; liver tissue engineering, LTE.

Collagen is a significant constituent of the natural ECM and it consists of at least 19 subtypes that provide various functions. Collagen is naturally degraded by metalloproteases, specifically collagenase and serine proteases [8]. As a major determinant of the architecture and tensile strength of many tissues, collagen participates in numerous physiologically important interactions and was made into scaffolds, which have been used in a variety of applications due to a number of useful properties such as hemostatic, low antigenicity, and appropriate mechanical properties [103]. Collagen and glycosaminoglycans compose a considerable portion of the ECM to ensure the mechanical integrity of hepatocytes and are responsible for providing bioactive molecular signals to cells [112]. Platelet deposition and hepatocyte culture experiments showed that a new collagen/chitosan hydrogel had excellent blood and cell compatibility, which suggests that this hydrogel is a promising implantable candidate for LTE [79]. Andrea et al. optimized the collagen type I-hyaluronan hybrid hydrogel for liver microenvironments, which was employed to bioprint 3D liver tissue constructs containing primary human hepatocytes and liver stellate cells [113]. Similarly, collagen has been incorporated with other materials such as chitosan and heparin in order to recapitulate liver functions [80,114,115].

Gelatin is a protein produced by partial denaturalization or hydrolytic degradation of collagen and has a sol-gel transition temperature around 30 °C [104]. Due to its natural origin, gelatin possesses biological activities and has a high ability to form strong hydrogels and transparent films that are easily designed as insoluble hydrophilic polymers. Gelatin induced essential cellular functions, such as migration, proliferation and differentiation through integrin-mediated cell adhesion and cell-mediated enzymatic degradation [105]. Using rapid prototyping technology, hepatocytes were laminated into gelatin hydrogels for more than 30 layers, remained viable, and performed biological functions in the construct for more than two months [116]. More interestingly, a heparin–gelatin mixture was used to coat the vasculature within decellularized livers to reconstruct a patent vascular tree by seeding endothelial cells [117]. To make a whole bioengineered liver, gelatin was incorporated with polyurethane to generate a hydrogel with controlled pore size and interconnectivity for LTE [118].

Hyaluronan (hyaluronic acid, HA) is a non-sulphated glycosaminoglycan consisting of alternating units of D-glucuronic acid and D-N-acetylglucosamine, which are linked via beta-1,4 and beta-1,3 glycosidic bonds [99]. As one of the major components of ECM, HA is naturally degraded by hyaluronidase allowing cells in the body to regulate the clearance of the material in a localized manner. Unmodified HA binds to water and promotes swelling of the matrix and additionally can inhibit cell-cell adhesion by forming a porous coat around cells [119]. HA can also provide signals to enhance cell attachment and migration once modified with appropriate cell-adhesive proteins and peptides [104,120]. Therefore, HA has been used extensively for LTE applications. HA hydrogels used to be identified as the only culture condition that facilitated survival, proliferation and maintenance of hepatoblasts and could support human liver cells, including several subpopulations of hepatic progenitors [121]. Recently, Jonas et al. successfully cultured hepatocytes in a liver-on-a-chip setup by using a modular hyaluronan-PEG based 3D hydrogel modified with RGD peptides [122].

Fibrin can be isolated autologously from patients and fabricated into hydrogel scaffolds. Actually, fibrin was first noted to have a hemostatic effect on wounds and was subsequently applied to cerebral hemorrhage. With refinements adding to the strength, efficacy and safety, fibrin glues have become a more popular tool in the application of tissue-engineered skin replacements [109]. As fibrin can achieve high seeding efficiency and uniform cell distribution [110], fibrin hydrogels have also been utilized for LTE. Helge et al. evaluated a fibrin-based hydrogel and found it suitable for the stimulation of hepatocytes and it appeared to support engraftment and specific differentiation of viable hepatocytes [123]. Fibrin hydrogels together with PLGA and hepatocytes were assembled to an implantable liver tissue, along with a hierarchical vascular network [124]. Most recently, a fibrin hydrogel was successfully utilized for the ectopic expansion of engineered human liver tissue using mature cell populations [125].

Alginate is a polymer consisting of beta-D-mannuronic acid (M) and its alpha-L-glucuronic acid (G), and it is commonly found in the cell wall of brown seaweed and produced extracellularly by some bacteria [101,104]. As an anionic polysaccharide, alginate can easily create hydrogels in the presence of divalent cations and can mimic the ECM well, which makes it popular for LTE. One of the challenges in fabricating liver in vitro is the inability to culture hepatocytes. Using alginate-based scaffolds, hepatocytes were successfully cultured for two weeks while maintaining the hepatocyte phenotype [126]. Hence, scaffolds fabricated by 3D printing hold new promise in creating functional liver tissues [127]. Recently, an injectable hydrogel made from glycyrrhizin (GL), alginate (Alg), and calcium (Ca) was designed for application in LTE, and the GL–Alg–Ca hydrogel could maintain proliferation and liver specific functions of a hepatic cell line [128].

Chitosan is derived from the deacetylation of chitin, which is a linear polysaccharide consisting of beta-1,4 linked N-acetylglucosamine units. Chitin is the most abundant natural biopolymer besides cellulose and has highly hydrophobic and electric properties. Different from chitin, chitosan is a soluble polymer with high biofunctionality and better adsorption. Chitosan is capable of cell adherence and proliferation, and taking its ability to form highly porous scaffolds and antibacterial properties into consideration, chitosan is a promising choice for LTE. Pure chitosan-based microfibers were prepared to support self-aggregation of liver cells into spheroids, showing improved liver specific functions [129]. He et al. made use of well-organized microstructures for hepatic tissue engineering with chitosan-gelatin hybrid scaffolds [130]. Furthermore, with the fibronectin coating on the surface, the chitosan nanofibers exhibited a significantly enhanced cell attachment and the hepatocytes in co-cultures formed colonies and maintained their morphologies and functions for prolonged periods of time [131].

Polyhydroxyalkanoates (PHAs) are a group of aliphatic polyesters synthesized by bacteria to store intracellular carbon and energy, including more than 150 identified monomers [29]. Various monomers provide a broad range of properties to engineer multifunctional constructs that have poor stiffness and slow degradation rate. Su et al. [132] developed scaffolds for LTE using poly(3-hydroxybutyrate-co-3-hydroxyvalerate-co-3-hydroxyhexanoate) (PHBVHHx). The matrices derived were loaded with human umbilical cord multipotent stromal cells (MSCs) and hepatocyte-like cells, and after 28 days the tissue generated looked very similar to the native organ. A study reported the recovery of injured mouse liver when a PHBVHHx scaffold loaded with human umbilical cord Wharton's jelly (WJ) MSCs was transplanted [133]. Chemically modified PHAs also find use as films, pins, sutures, screws, and scaffolds for repairing skin, cartilage and LTE [111].

Cellulose: In contrast with most other biopolymers, gelation of various cellulose derivatives including MC and hydroxypropylmethylcellulose (HPMC) occurs upon heating. Cellulose is often combined with proteins (e.g., gelatin), polysaccharides (e.g., chitosan), or both. Other cellulose derivatives have been reviewed by Vlierberghe, et al. [104]. Cellulose nanofibrils, which are fibrils in the nanometer range, show general properties of cellulose: hydrophilicity and broad chemical modification capacity combined with properties specific for nanoscale materials due to their high surface area. With good mechanical properties and biocompatibility, cellulose nanofibrils are attractive for biomedical applications [32,134]. Nanofibrillar cellulose hydrogel was shown to promote three-dimensional liver cell culture [82]. A hydrogel composed of alginate and cellulose nanocrystal was suitable for bioprinting of liver-mimetic honeycomb 3D structures [135]. Wood-derived nanofibrillar cellulose (NFC) has been incorporated with hyaluronan-gelatin (HG) to form hydrogels for the differentiation of liver progenitors, and undifferentiated progenitor cells in NFC-HG hydrogels formed 3D multicellular spheroids with apicobasal polarity and functional bile canaliculi-like structures, structural hallmarks of the liver tissue [136].

Agarose is a linear polysaccharide formed by the disaccharide of beta-D-galactose and 3,6-anhydro-a-L-galactopyranose. Agarose is extracted from seaweed and can be dissolved in hot water. It forms a gel upon cooling due to the formation of double helices and their subsequent aggregation. The thermo-reversible gelation process depends on the type of agarose or methoxy content [97]. Agarose gels have adjustable pore sizes and are physicochemically strong, which enables high diffusion rates. Primary hepatocytes could proliferate in vitro in an agarose-chitosan scaffold, with suitable physicochemical properties and hepatic cell compatibility, and showed an increase in cellular metabolic activity. Hepatic functions like albumin secretion and urea synthesis were improved for primary hepatocytes in the 3D scaffold compared to controls [97].

3.2.2. Synthetic Hydrogels

Synthetic hydrogels are artificial hydrogels with a defined composition and structure. Compared to biological hydrogels, synthetic hydrogels are less complex and have stronger mechanical structure, less animal origination, are well controlled, commercially friendly, and relatively easier to be FDA-approved, which makes them more and more popular. Several synthetic materials utilized to make hydrogels are introduced in the following section and Table 2.

Table 2. Characteristics and applications of synthetic materials.

Synthetic Materials	Major Concerned Properties				Other Characteristics	Applications	References
	Biocompatible	Biodegradable	Hydrophilic	Thermal-Responsive	FDA			

Synthetic Materials	Biocompatible	Biodegradable	Hydrophilic	Thermal-Responsive	FDA	Other Characteristics	Applications	References
PAA	Yes	No	Yes	No	Approved	Various derivatives, excellent optical transparency and stability in water	Degradable or thermal-responsive derivatives for drug delivery	[31]
PVA	Yes	No	Yes	Yes	Approved	Inefficient elasticity, stiff membrane, lack of cell adhesion, ease of fabrication and sterilization	Tissue engineering, both medical and nonmedical devices	[36,137]
PIC	Yes	No	Yes	Yes	No	Semiflexible properties, strain stiffening	Tissue engineering and cancer immunotherapy	[138,139]
PEG	Yes	No	Yes	No	Approved	Hydrolytically bioactive, photocrosslinkable, easily modifiable	Widely used in for chemical, biological, and commercial purposes, and also in tissue engineering	[31,140]
PLGA	Yes	Yes, controlled	No	Yes	Approved	Poor load-bearing properties, hydrolytically unstable; good cell adhesion and proliferation	Medical devices, drug delivery, fabrication of tissue engineering matrices, suture reinforcements	[141,142]
PGA	Yes	Yes	Yes	No	Approved	Highly crystalline, high melting point, lacks elasticity, not soluble in most organic solvents, tends to lose mechanical strength	Absorbable sutures, orthopedic devices, scaffolding matrices for tissue regeneration	[81,143]
PLA	Yes	Yes, slow	No	No	Approved	Lacks elasticity, high tensile strength, hydrolytically unstable	Orthopedic fixation devices	[143,144]
PCL	Yes, less	Yes. Low rate	No	Yes	Approved	Limited elasticity, tunable mechanical properties	Tissue engineering, long-term drug/vaccine delivery vehicle	[35,145]

Abbreviations: poly(ethylene glycol), PEG; poly(lactide-co-glycolide) acid, PLGA; polyisocyano peptide, PIC; Poly(vinyl alcohol), PVA; poly(N-vinylpyrrolid), PVP; poly(propylene furmarate-co-ethylene glycol), P(PF-co-EG); poly(2-hydroxyethylmethacrylate), HEMA; poly(acrylic acid), PAA; polyglycolic acid, PGA; polylactic acid, PLA; poly-ε-caprolactone, PCL; poly(N-isopropylacrylamide), PNIPAAm.

Poly(ethylene glycol) (PEG) is a polyether compound which is water-soluble, amphiphilic, transparent, colorless, liquid, and viscous. Various modifications have been applied to PEG to enhance the mechanical properties for 3D printing, to contribute to high elasticity, or to increase hydrophilicity which could tune the degradation rate [78]. PEG derivatives were used as crosslinkers to develop bioartificial vessel-like grafts [146]. Nowadays, it has become a frequently employed strategy to increase protein solubility and stability to reduce immunogenicity and to alter circulation half-life [139]. For LTE, PEG hydrogels are widely used for encapsulation, and was shown to provide a biocompatible matrix that allows the majority of encapsulated primary hepatocytes to survive [147]. The survival and function of PEG hydrogel-encapsulated hepatic cells have been improved by modifications in polymer chain length and the conjugation of bioactive factors [8]. Moreover, hepatic cells have been encapsulated well into the photopolymerized PEG hydrogel through which complex architecture constructs were assembled [148]. The undegradable PEG hydrogel was applied for the encapsulation of co-cultured hepatocytes, preventing aggregation and overgrowth, and enabling formation of microtissues with stable hepatic function [149]. Recently, PEG was fabricated into 3D hexagonally arrayed lobular human liver tissues and the hydrogel enabled primary human fetal liver cells to self-assemble into a 3D configuration and preserved advanced hepatic functions for at least five months [150].

Polyisocyanopeptide (PIC) is an innovative and fully synthetic polymer, capable of mimicking characteristics of the natural ECM [139]. PIC exhibits thermo-reversible behavior due to the hydrophobic interactions of the oligoglycol substituent present along its backbone, with a steep increase of the storage modulus (G') above 18 °C. As a water-soluble synthetic polymer, PIC mimics natural protein-based filaments. Its thermoreversible gelation property and cytocompatibility make PIC an ideal candidate for bioprinting technology [151]. The unique semiflexible properties combined with a length of several hundred nanometers have recently made it particularly attractive for LTE [139].

Poly(vinyl alcohol) (PVA) is prepared in two steps due to the unstable form of vinyl alcohol as monomeric units. By controlling the hydrolysis step, different grades of PVA polymer can be prepared, which finally affects the behavior of the polymer material, solubility, crystallinity, and chemical properties [152]. PVA-based hydrogels have been applied to many kinds of tissues, such as skin, bone, cartilage, vascular- and cardiac-tissue, human prostate and artificial cornea. Due to its favorable properties and easy manipulation, PVA-based hydrogels have been recognized as promising biomaterials and are suitable candidates for LTE. To overcome disadvantages such as poor cell-adhesion, they still need further modifications for targeted applications [36]. Shan et al. developed a method to prepare transparent PVA hydrogels by varying the freeze/thaw cycles and the PVA hydrogels exhibited similar mechanical properties and morphological characteristics to that of a porcine liver, a reference material for human soft tissue [137]. PVA/gelatin hydrogels were proposed as a 3D microenvironment for liver cells to form an in vitro hepatocellular carcinoma model [153].

Poly(lactide-co-glycolide) acid (PLGA) is an FDA approved biodegradable material and has been studied widely both in vivo and in vitro. Previous studies have shown poor load bearing properties [141]. Due to its biocompatibility and controllable biodegradability, PLGA microspheres have been utilized as scaffolds containing cells to enhance the vascularization of engineered tissues. Besides, PLGA is also attractive for its property to be degraded by hydrolysis to lactic acid and glycolic acid [142]. Therefore, PLGA hydrogels have been used frequently for LTE. More than 20 years ago, PLGA was fabricated into scaffolds for LTE and seeded with hepatocytes and non-parenchymal cells from rats [154]. When cultured together with biodegradable PLGA membranes, the cells in the 3D stacked structures recovered polarity and exhibited improved liver-specific functions as compared with cells in a monolayer [155]. Moreover, the transdifferentiation rates of bone marrow mesenchymal stem cells (BMSCs) to mature hepatocytes were improved by collagen-coated PLGA [84]. Recently, PLGA polymer has been utilized to fabricate an absorbable vascular anastomosis device and the device was tested in pig liver transplantation experiments, where it was successfully absorbed within four months [83].

Poly(glycolide) acid (PGA) is the simplest linear aliphatic polyester and was used to develop the first totally synthetic absorbable suture. With a high degree of crystallization and high melting point, PGA is not soluble in most organic solvents except for the highly fluorinated ones. As an absorbable material, its thermal stability is good. Unfortunately, PGA tends to lose its mechanical strength rapidly due to the hydrophilic nature. Sutures of PGA will lose around 50% of their strength after two weeks and 100% at four weeks, and will get completely absorbed in 4–6 months [143]. With this property, a PGA felt was incorporated with fibrin sealant for prevention of bile leakage after liver resection [156].

Poly(lactide) (PLA) lactide is the cyclic dimer of lactic acid with two optical isomers. L-lactide is the naturally occurring isomer and DL-lactide is the synthetic blend of D- and L-lactide. The polymerization of lactide is similar to that of PGA. With a pendant methyl group on the alpha carbon, PLAs are quite different in chemical, physical and mechanical properties when compared to PGA, even though their structures are similar, and PLA is more frequently utilized in LTE. Rat hepatocytes cocultured with primary rat hepatic stellate cells on the PLA hydrogels have been shown to maintain hepato-specific functions for more than two months [157]. The biodegradable copolymer poly(lactic acid-co-lysine) (PLAL) contributed to hepatocyte engraftment, function and expansion [158]. Type I collagen coated electrospun poly(L-lactic acid) (PLLA) nanofibers with random and aligned orientation were evaluated for hepatocyte adhesion and proliferation [159]. PLLA and gelatin were used to induce hepatic differentiation of MSCs in the form of electrospun nanofiber scaffolds and the microporous scaffolds controlled the migration of hepatic stellate cells through pore size [9].

Poly(e-caprolactone) (PCL) is an aliphatic polyester with a glass transition temperature at −60 °C. The ring-opening polymerization of e-caprolactone yields a semi crystalline polymer and gives softness and flexibility at near body temperature. This polymer has been regarded as tissue-compatible and used as a biodegradable suture in Europe. Furthermore, the very low degradation rate makes it suitable for long-term implants or for drug delivery systems [148]. PCL combinations with a variety of natural polymers were reported for LTE [145]. PCL has been used to enhance mechanical properties and could be bioprinted together with hepatocytes, endothelial cells and fibroblasts, which maintained hepatocyte functions and facilitated the formation of vascular networks [160]. Rhiannon et al. developed hybrid PCL-ECM scaffolds for LTE, which maintained hepatocyte growth and function [161]. In addition, PCL nanofiber scaffolds supported the in vitro differentiation of human somatic stem cells into hepatocytes [162]. Besides hepatocytes, PCL/chitosan electrospun nanofibers were evaluated to be competent for the culture of mouse hepatic cells, indicating that PCL/chitosan hydrogels would be excellent for LTE [163].

Poly(acrylic acid) (PAA) and its derivatives are among the most intensively studied synthetic materials for biomedical applications. Several attempts have been made for their application in LTE. When grafted within PAA, the growth kinetics of adhesion patch at primary hepatocyte cell substrate interface was changed [164]. Amol et al. [165] conjugated PAA and polyethyleneimine (PEI) with elastin-like polypeptides (ELPs) and found that the conjugates influenced the morphology, aggregation and differentiation function of primary rat hepatocytes.

3.3. Progress in Hydrogel Techniques

Great progress has been achieved in hydrogel techniques to provide as many cues as possible for mimicking the ECM. As mentioned above, various design strategies to overcome the shortcomings of individual biomaterials were developed and many different hydrogels that successfully mimic the complexity of natural ECMs have been created.

The aim of different design strategies is simply to make the best use of ideal characteristics of various biomaterials and recapitulate as many ECM functions as possible. The principle to design those hydrogels is based on the properties required by the liver. Up till recently, properties such as biocompatibility, biodegradability, adhesive property, thermal-responsiveness and purposed stiffness and swelling have come into being via elaborate designs.

Multidisciplinary design optimization has been tried to devise ideal hydrogels (Table 3). Several aspects have been studied in great detail, such as gel formation dynamics, crosslinking modes, and mechanical and degradable material linkages. These properties are linked to the intrinsic properties of the main chain polymer and the crosslinking characteristics (amount, type, and size of crosslinking molecules) [8]. The most common types of crosslink include: covalent, physical, dynamic covalent, hydrogen bonding, affinity bonding, hydrophobic interaction, and chain entanglement. Compared to chemical means, which may be toxic and could affect the nature of substances, physical mechanisms are safer. Different from conventional chemical or physical methods, photopolymerization seems more promising and is attracting more attention.

Methods mentioned above provide various possibilities for hydrogel design [140]. Hydrogel compositions can be reinforced by polymers or other hydrogels. To promote cell adhesion, peptides and fragments are used. Polymeric hydrogel adhesives could be synthesized by physical or chemical gelation or by the combination of both. Different materials (polysaccharide-/protein-/or synthetic polymer-) based hydrogel adhesives own quite different characteristics, signaling properties included. For example, the design of galactose-carrying hydrogels as ECMs can guide hepatocyte adhesion and enhance cell functions [166]. Another highly studied property is the degradability of hydrogels. At present, hydrolysis and enzymatic methods are still the main strategies for hydrogel degradation.

With the combination of multiple design strategies, hydrogels tend to gain more comprehensive properties and methods to characterize properties of hydrogels have also been increasing. Up till now, the most frequently characterized properties are gelation time, gel fraction, swelling degree, structural parameters, water vapor transmission rate, and mechanical properties. Smart hydrogels with high tunability of stiffness can be designed with various modifications, which enable hydrogels to be pH-/ thermo-/ photo-/ redox-, or mechano-responsive [92].

Bioengineering **2019**, 6, 59

Table 3. Hydrogels designed for liver tissue engineering.

Composition	Cell Source	Crosslinking Method	Output	Reference
Collagen, chitosan	Platelet and rat hepatocyte; rat hepatocyte	Chemical crosslinking; noncovalently linked	The matrix has excellent blood and cell compatibility; hepatocytes exhibited relatively high glutamate-oxaloacetate transaminase and glucose secretion functions	[79,114]
Collagen, chitosan, heparin	Platelet and rat hepatocyte	Chemical and physical crosslinking	Improved the blood compatibility and maintained hepatocyte viability and function; exhibited high urea and triglyceride secretion functions	[80,113,115]
Collagen I, HA	Primary human hepatocytes and liver stellate cells	Physical crosslinking, UV crosslinker	Bioprinted 3D liver tissue constructs maintained liver functions including urea and albumin production	[113]
Gelatin, chitosan	Human HepG2; primary rat hepatocyte	Crosslinked with 1% genipin; crosslinked by glutaraldehyde solution	Cells cultured in 3D scaffolds preformed better on the structural characteristics, cell viability, growth and liver specific functions; supply living cells with nutrients and allow removing the cell metabolite; hepatocytes perform better in the well-defined scaffold	[167–169]
Gelatin, silk fibroin (SF)	Human normal hepatic QZG cell line	Use of glutaraldehyde solution to produce cross-linked gelatin solution and then mix with SF	Achieved better biocompatibility, controlled degradation, and good for the attachment and proliferation of cells	[170]
Gelatin	Primary rat hepatocytes	Gelatin is dissolved in hot NaCl and Tris-HCl	Rapid prototypedg hepatocytes remained viable and performed biological functions for more than 2 months	[116]
Gelatin, heparin	Human endothelial cells and HepG2 cells	Physical mixture	Helped cells to reconstruct a patent vascular tree within the decellularized porcine liver scaffold	[117]
Gelatin, polyurethane	Hepatocyte	Cross-linked with glutaraldehyde, enhanced by the addition of lysine	Generation of a hydrogel with controlled pore size and interconnectivity	[118]
GelMA	Human HepG2/C3A cells	Photocrosslinked	Bioprinted liver spheroids exhibited long-term functionality	[112]
HA, PEG	Human HepG2 cells, hiPS-HEPs	Bioorthogonal SPAAC crosslinked, modified with cyclic RGD peptides	hiPS-HEPs migrated and grew in 3D and showed an increased viability and higher albumin production compared to ctrols	[122]
HA, moieties; collagen III, laminin	Primary rat hepatocytes; hHiPSCs	Galactose moieties were covalently coupled with HA through ethylenediamine; the is initiated by a PEGDA cross-linker	Formation of cellular aggregates with enhanced liver specific metabolic activities and improved cell density; permissive for survival and phenotypic stability of human hepatic stem cells and hepatoblasts	[171,172]

Table 3. *Cont.*

Material	Cells	Method	Description	Ref.
Fibrin	Rat hepatocyte; human hepatocytes, dermal fibroblasts, and UVECs	Human fibrinogen was applied with the thrombin solution to make the fibrin matrix	Supported engraftment and specific differentiation of viable hepatocytes; stimulated hepatocytes for the ectopic expansion of engineered human liver tissue seeds; in vitro-generated liver tissues can expand and function in vivo	[110,123,125]
Fibrin, PLGA	Rat hepatocytes and ADSCs	Formed by the pollymerization of fibrinogen acted by protease thrombin	Assembled to be an implantable endothelialized liver tissue, along with a hierarchical vascular network	[124]
Alginate	Mouse primary hepatocytes; HepG2 cells	Freezedry technique; crosslinked in CaCl2 solution	Maintained hepatocyte genotype, produced hepatic-specific proteins for two weeks; liver spheroids displayed an enhanced cell proliferation; importance of cell density within weakly adhesive alginate scaffolds; a cold reduction in temperature display an enhanced cell proliferation	[101,126,127,173,174]
Alginate, galactosylated chitosan	Primary hepatocytes	Calcium crosslinked; lyophilization	Enhanced hepatocyte aggregation; improved cell attachment and viability	[175,176]
Alginate (Alg), glycyrrhizin (GL), calcium (Ca)	HepG2 cells	Calcium crosslinked equal volume mixture of GL, nano-CaCO3 and Alg	GL–Alg–Ca hydrogel was homogenous complex with stable structure and well viscoelasticity, and cells showed good biocompatibility, and maintained the viability, proliferation and liver function	[128]
Chitosan	HepG2 cells	The microfluidic fabrication process for pure chitosan microfibers	HepG2 cells were self-aggregated with a spheroid shape, showing a higher liver specific function (albumin secretion and urea synthesis).	[129]
Lactose-modified chitosan (Lact-CTS)	Normal liver cell	Coupling of lactose with chitosan was carried out by the reducing agent, addition of NaBH4	Lact-CTS with 48.62% of galactose moieties could facilitate the cell attachment and possess great biocompatibility and mechanical stability	[177]
Chitosan, gelatin	Hepatoytes	Crosslinked by glutaraldehyde solution	Scaffold produced with predefined multilevel internal architectures (a flow-channel network and hepatic chambers) and improved hepatocytes performance greatly in comparison with a porous scaffold	[130]
Silk fibroin/chitosan (SFCS)	HepG2 cells	Freezing and lyophilization	Provided a matrix with homogeneous porous structure, controllable pore size and mechanical properties	[178,179]
Chitosan nanofibers, fibronectin	Primary rat hepatocytes, endothelial cells	Fabricated by the electrospinning technique	Enhanced cell attachment and maintained their morphologies and functions	[131]
PHBVHHx	UC-MSCs, hepatocyte cells	Solid–liquid phase separation method to form scaffolds	Injured mice liver were recovered; generated tissue looked similar to the organ	[132,133]

Table 3. *Cont.*

Material	Cells	Crosslinking	Description	Ref
Native nanofibrillar cellulose (NFC)	Human hepatic cell lines HepaRG and HepG2	Physically crosslinked	Provided mechanical support for cell growth and differentiation, and induced spheroid formation of HepaRG and HepG2 cells.	[82]
Cellulose nanocrystral (CNCs), alginate	Human hepatoma cells, fibroblasts	Crosslinked with CaCl2	The bioink formulation was suitable to print a liver mimetic honeycomb 3D structure containing fibroblast and heptatoma cells	[135]
Nanofibrillar cellulose, HA-gelatin	Human HepaRG liver progenitor cells,	HG hydrogel based on thiol-modified HA, thiol-modified gelatin and crosslinker PEGDA	Induced apicobasal polarity and functional bile canaliculi-like structures, expediting the hepatic differentiation of HepaRG liver progenitor cells better than the standard 2D culture	[136]
Agarose, carbohydrate glass	Primary rat hepatocytes and fibroblasts	Chain entanglements, physical crosslinking	Primary hepatocytes and fibroblasts were cast	[180,181]
Agarose–chitosan (AG–CH)	Primary rat hepatocytes	Crosslinke by glutaraldehyde	The hepatic functions like albumin secretion and urea synthesis were established in the 3D scaffold	[97]
PEG, heparin	Primary rat hepatocytes, BMEL; cryopreserved primary human hepatocytes, induced pluripotent stem cells (iPSCs)	UV light polymerization; chemically crosslinked	Demonstrated the importance of cell–cell and cell–matrix interactions in BMEL cell and primary hepatocyte survival. Aggregation and encapsulation of iPS cells during their differentiation towards hepatocytes yielded microtissues that depicted stable albumin production on-chip and inducible CYP activity. The 3D in vitro liver model is capable of sustaining advanced human-specific liver functions	[147–149, 182,183]
PEG, PLGA, liver-derived ECM (L-ECM), growth factors	Rat liver	Physical and thermal crosslinking	L-ECM and growth factors enhanced tissue penetration into intrahepatically implanted biodegradable scaffolds and induced cell proliferation in the parenchyma that surrounds these scaffolds in the normal liver	[184]
PEG-DA, PEGDAAm, MMP-sensitive	Primary human fetal liver cells, HUVECs and HepG2	Chemical crosslinking, photopolymerization	The 3D in vitro liver model is capable of sustaining advanced human-specific liver functions for at least 5 months in culture. Hepatic tissues survived and functioned for over 3 weeks after implantation	[150,185]
PIC, GRGDS peptide	Human dermal microvascular endothelial cells and fibroblasts	Polymerization of the corresponding monomers using a nickel perchlorate as a catalyst	Supported pre-vascularization and the development of organotypic structures	[186]
PVA		Physical crosslinking by freeze-thaw cycless	Exhibited similar mechanical properties and morphological characteristics to porcine liver	[137]

Table 3. Cont.

Material	Cells	Crosslinking	Outcome	Ref.
PVA, gelatin	HUVECs and HepG2	Physically crosslinked by freeze–thaw cycles	Hydrogel particles with a well pronounced tendency towards association with hepatocytes and endothelial cells.	[152]
PLGA, AVAD; collagen-coated PLGA (C-PLGA); PLGA, gelatin	Pig liver transplantation; rat BMSCs; rat hepatocytes (HCs), nonparenchymal liver cells	Chemical and physical crosslinking	Demonstrated the feasibility of using AVADs in organ transplantation. Proved the superiority of the C-PLGA for hepatocytes differentiation. HCs cocultured with nonparenchymal cells can attach to and survive on the 3D polymer scaffolds. Cells recovered polarity and exhibited improved liver-specific function. Untreated PLGA performed best for supporting liver-specific functions. 3D printing and optimized parameters are applied for liver regeneration.	[83,84,154,155,187,188]
PGA, fibrin		Physical crosslinking	Effective in preventing biliary leakage	[156]
PLA	Rat hepatocytes, rat hepatic stellate cells	Dissolved in 2,2,2-trifluoroethanol	Encouraged the rapid self-organization of 3D spheroids and the spheroids formed exhibit hepatocyte-specific functionality	[157]
PLAL	Rat hepatocyte	Chemical crosslinking	Contributed to hepatocyte engraftment, function, and expansion	[158]
PLA, fibroin; collagen I; discrete aligned nanofibers	HepG2 cells; rat hepatocyte	Chemical and thermal crosslinking	Improved the cell growth, enhancing cells adhesion and proliferation. Hepatocyte aggregates formed on nanofibers displayed excellent cell retention, cell activity and stable functional expression	[144,159]
PLLA, gelatin, HGF	BMSCs	Electrospinning, physical and chemical crosslinking	Effectively guide hepatic commitment of patient derived BMSCs	[9,189]
PCL, collagen	Primary rat hepatocytes, HUVECs and human lung fibroblasts (HLFs)	Physical and thermal crosslinking, 3D printing	The vascular formation and functional abilities of HCs demonstrated that the heterotypic interaction among HCs and nonparenchymal cells increased the survivability and functionality of HCs	[160]
PCL, ECM	HepG2 hepatocytes	Electrospinning, physical and thermal crosslinking	Provided a viable, translatable platform for hepatocytes; supporting in vivo phenotype and function	[161]
PCL	Human USSCs, self-renewing pluripotent cells	Physical crosslinking, electrospinning	Differentiation of USSCs demonstrated that this culture system can potentially be used as an alternative to the ECM-based culture for relevant hepatocyte-based applications in LTE	[162]

Table 3. *Cont.*

PCL, chitosan	Epithelial liver mouse cells.	Physical crosslinking, electrospinning	The porosity and pore is suitable for epithelial liver mouse cells infiltration, attachment, and material exchange	[163]
PAA, PET, collagen	Primary hepatocytes	Chemical an physical crosslinking, UV light induced polymerization	The growth kinetics of adhesion patch at primary hepatocyte cell substrate interface is changed upon PAA grafting	[164]
PAA, PEI, ELPs	Primary rat hepatocytes,	Chemical an physical crosslinking	ELP–polyelectrolyte conjugates profoundly influenced the morphology, aggregation and differentiation function of primary rat hepatocytes	[165]

Abbreviations: hyaluronic acid, HA; polyhydroxyalkanoates, PHAs; poly(ethylene glycol), PEG; poly(lactide-co-glycolide) acid, PLGA; polyisocyano peptide, PIC; poly(vinyl alcohol), PVA; poly(N-vinylpyrrolid), PVP; poly(propylene furmarate-co-ethylene glycol), P(PF-co-EG); felatin methacryloyl, GelMA;poly(acrylic acid), PAA; poly(ethylene terephthalate), PET; polyethyleneimine, PEI; polyglycolic acid, PGA; polylactic acid, PLA; poly-e-caprolactone, PCL; poly(N-isopropylacrylamide), PNIPAAm; hepatoma cells, HepG2; strain-promoted alkyne-azide 1,3-dipolar cycloaddition, SPAAC; Arg-Gly-Asp, RGD; bone marrow mesenchymal stem cells, BMSCs; matrixmetalloproteinase sensitive peptide, MMP-sensitive; human iPSC derived hepatocytes, hiPS-HEPs; human hepatic stem cells, hHpSCs; (human) umbilical vein endothelial cells, (H)UVECs; adipose-derived stem cells, ADSCs; bipotential mouse embryonic liver cells, BMEL; absorbable vascular anastomotic device, AVAD; elastin-like polypeptides, ELPs; human cord blood–derived unrestricted somatic stem cells, USSCs.

4. Conclusions

Taking the biological and physicochemical properties into consideration, the characteristics that are significant for LTE are summarized (Figure 2a) and some specific properties are suggested (Figure 2b), which may facilitate the choice for a specific hydrogel to mimic ECM for LTE.

In view of the biological origin of natural materials, the majority is biocompatible, biodegradable, and abundantly available. As most of these natural materials are present in ECM, cells have a good compatibility and growth response. Being more bioactive compared to synthetic hydrogels, natural hydrogels have a longer history of research as well as more utilization in TERM, especially since several of them have been FDA-approved. However, every coin has two sides, and this is also true with regard to natural hydrogels. Compared with synthetic materials, natural hydrogels have several shortcomings, such as mechanical weakness, batch-to-batch variability, and the fact that some are animal-derived, which implies ethical issues and restricts the utility for clinical applications. Obviously these drawbacks do not, or to a lesser degree, account for synthetic hydrogels.

Synthetic hydrogels are either modified from natural materials or completely synthetic, and based on the type of material, synthetic hydrogels offer choices to be degradable or nondegradable. Compared to natural hydrogels, synthetic hydrogels are relatively less immunogenic, quality-reproducible, mechanically stronger, and easily modifiable [158]. Strong mechanical properties and various modifications have increased the popularity of synthetic hydrogels in TERM. However, synthetic hydrogels are still far from perfect for LTE, and the most significant weakness is that many of them are less bioactive and lack viscoelasticity.

As neither natural-nor synthetic-hydrogels alone are suitable for LTE, the combination of different hydrogels, with different origins or various modifications, has been applied in LTE, and great progress has been achieved in the past decades. As an example, the overall performance of the liver cell-loaded PCL scaffolds was remarkably improved by avidin–biotin binding-based cell seeding [190]. In addition, it has been shown that presence of poly(3,4-ethylenedioxythiophene) (PEDOT) as a conducting polymer in the scaffolds, with the combination of gelatin/ chitosan/ hyaluronan, enhanced hepatocyte cell viability, attachment and proliferation [191].

(a)

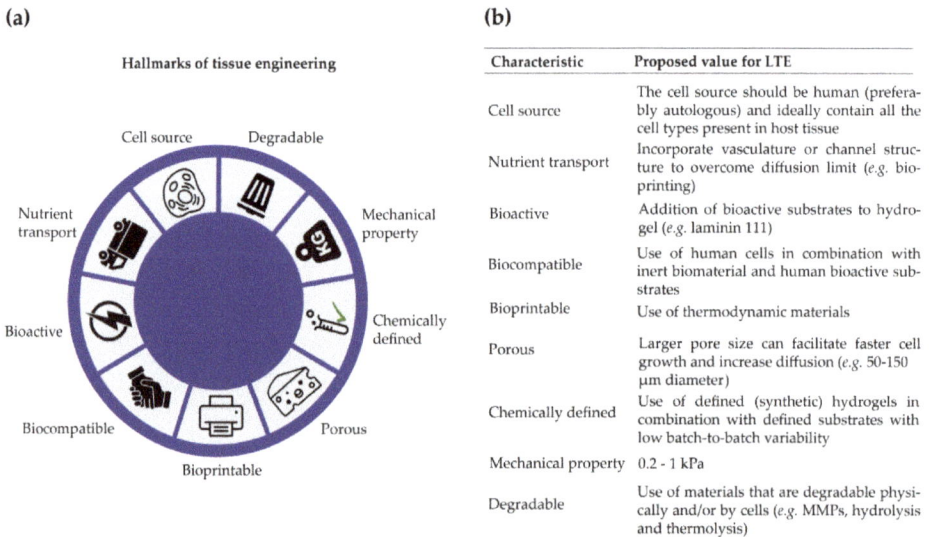

Hallmarks of tissue engineering

(b)

Characteristic	Proposed value for LTE
Cell source	The cell source should be human (preferably autologous) and ideally contain all the cell types present in host tissue
Nutrient transport	Incorporate vasculature or channel structure to overcome diffusion limit (*e.g.* bioprinting)
Bioactive	Addition of bioactive substrates to hydrogel (*e.g.* laminin 111)
Biocompatible	Use of human cells in combination with inert biomaterial and human bioactive substrates
Bioprintable	Use of thermodynamic materials
Porous	Larger pore size can facilitate faster cell growth and increase diffusion (*e.g.* 50-150 μm diameter)
Chemically defined	Use of defined (synthetic) hydrogels in combination with defined substrates with low batch-to-batch variability
Mechanical property	0.2 - 1 kPa
Degradable	Use of materials that are degradable physically and/or by cells (*e.g.* MMPs, hydrolysis and thermolysis)

Figure 2. Hydrogels for liver tissue engineering (LTE). (**a**) Hallmarks of tissue engineering; (**b**) Proposed value of characteristics specific for LTE.

5. Discussion

Constructing a physiologically relevant bioengineered liver is of great interest as an in vitro model for fundamental and applied research such as disease pathogenesis, drug metabolism, and toxicological studies. Moreover, building physiologically relevant models now for in vitro studies will at the same time enhance our knowledge and progress towards LTE for clinical applications such as OLT in the future. Hydrogels are one of the most vital ingredients besides cells for bioengineered livers. Although great progress has been made in the past decades, there are still several major issues to be taken into consideration.

First of all, multiple liver cell types should be included to make a more physiologically relevant liver and characteristics required for hydrogels may vary among different cell types. In addition, bioengineering technologies have to allow the spatial orientation of these hydrogels in order to be planted at different positions to form microstructures. For example, viability and hepatic cell function were improved in micropatterned constructs as compared to unpatterned controls, demonstrating the importance of recreating the native microarchitectural features [91].

Secondly, there is a necessity to gain comprehensive understanding of liver ECMs. Liver ECM takes up 16–22% of the total liver volume [68,192,193], and is composed of various cues that can be divided into three categories [85]: supportive structure made from insoluble hydrated macromolecules (e.g., fibrillar proteins, proteoglycans, or polymer chains), soluble molecules (e.g., growth factors or cytokines), and noncellular factors (e.g., pH, temperature, charge). All those ECM effectors are possible determinants for the cell fate, interaction among cells, and the structure and function of tissues or organs. Similarly, liver cells can respond differently to various ECM components. Moreover, the ECM composition also varies in different parts of the liver (Figure 1c), which makes mimicking ECM for LTE more challenging. Therefore, several groups tried to use decellularized liver ECM as bioink for 3D cell-printing based LTE [194,195]. Nevertheless, the undefined chemical components of decellularized liver ECM will also restrict future applications in clinical treatment. Therefore, chemically defined hydrogels are still more promising for LTE. As more cues from liver specific ECM will be discovered, especially for the ECM within the space of Disse and the sinusoidal lumen, synthetic hydrogels will be able to mimic the in vivo microenvironment in much more detail.

Selection and design of hydrogels has to be carefully considered, and might differ depending on different applications [196]. To closer mimic the natural liver ECM, more details need to be included, which sets various strict requirements for hydrogels (Figure 2a). These requirements include: gel formation dynamics, crosslinking modes, biological and physicochemical properties, and degradable linkages. Importantly, the studies that are reviewed here and summarized in Table 2 did not only use different materials to mimic the ECM but also applied several different cell sources, cross-linking methods etc. This makes it difficult to directly compare the studies to each other, and to translate the outcome from one study to another. Nevertheless, some general conclusions on the requirements of hydrogels for LTE can be drawn, which are summarized in Figure 2b. For instance, the most dynamic effects of ECM stiffness on primary hepatocyte morphology and function were in the relatively narrow range between 150 Pa, the stiffness of normal liver, and 1 kPa, the lower threshold of fibrotic liver stiffness [87]. Primary hepatocytes demonstrated high viability and proliferation when seeded on 3D-printed gelatin scaffolds with precisely controlled pore geometry, and a physiologically mimetic 3D environment was proposed to be necessary to induce both expression and function of cultured hepatocytes [197]. Apart from those theoretical demands, several practical requests should also be kept in mind, especially for clinical applications. For instance, the hydrogels should be nonimmunogenic, easy to sterilize, and should enable engraftment post-implantation, being physically tunable to the in vivo microenvironment and the vascularization that has to be achieved within two days so that cells can survive and function.

In addition, related technologies have to keep up with the development of advanced hydrogels and their exquisitely designed characteristics, such as robust analysis technologies for local measurements of mechanical properties, and nanotechnology and bioprinting for promoting LTE.

Bioengineering **2019**, *6*, 59

Author Contributions: S.Y. and J.W.B.B. drafted the manuscript, K.S., B.S., and L.C.P. critically revised the manuscript for important intellectual content.

Funding: This research was funded by China Scholarship Council (CSC201808310180) to S.Y. and an NWO-VENI grant (016.Veni.198.021) to K.S.

Conflicts of Interest: The authors declare no conflict of interest.

References

1. Ananthanarayanan, A.; Narmada, B.C.; Mo, X.; McMillian, M.; Yu, H. Purpose-driven biomaterials research in liver-tissue engineering. *Trends Biotechnol.* **2011**, *29*, 110–118. [CrossRef] [PubMed]
2. Du, Y.; Han, R.; Ng, S.; Ni, J.; Sun, W.; Wohland, T.; Ong, S.H.; Kuleshova, L.; Yu, H. Identification and characterization of a novel prespheroid 3-dimensional hepatocyte monolayer on galactosylated substratum. *Tissue Eng.* **2007**, *13*, 1455–1468. [CrossRef] [PubMed]
3. Ng, S.; Han, R.; Chang, S.; Ni, J.; Hunziker, W.; Goryachev, A.B.; Ong, S.H.; Yu, H. Improved Hepatocyte Excretory Function by Immediate Presentation of Polarity Cues. *Tissue Eng.* **2006**, *12*, 2181–2191. [CrossRef] [PubMed]
4. Dash, A.; Inman, W.; Hoffmaster, K.; Sevidal, S.; Kelly, J.; Obach, R.S.; Griffith, L.G.; Tannenbaum, S.R. Liver tissue engineering in the evaluation of drug safety. *Expert Opin. Drug Metab. Toxicol.* **2009**, *5*, 1159–1174. [CrossRef] [PubMed]
5. Palakkan, A.A.; Hay, D.C.; Anil Kumar, P.R.; Kumary, T.V.; Ross, J.A. Liver tissue engineering and cell sources: Issues and challenges. *Liver Int.* **2013**, *33*, 666–676. [CrossRef] [PubMed]
6. Uygun, B.E.; Uygun, K.; Yarmush, M.L. 5.541—Liver Tissue Engineering. *Compr. Biomater.* **2011**, 575–585. [CrossRef]
7. Lee, J.S.; Cho, S.-W. Liver tissue engineering: Recent advances in the development of a bio-artificial liver. *Biotechnol. Bioprocess Eng.* **2012**, *17*, 427–438. [CrossRef]
8. Bhatia, S.N.; Underhill, G.H.; Zaret, K.S.; Fox, I.J. Cell and tissue engineering for liver disease. *Sci. Transl. Med.* **2014**, *6*, 1–21. [CrossRef]
9. Zhang, J.; Zhao, X.; Liang, L.; Li, J.; Demirci, U.; Wang, S. A decade of progress in liver regenerative medicine. *Biomaterials* **2018**, *157*, 161–176. [CrossRef]
10. Duan, B.-W.; Lu, S.-C.; Wang, M.-L.; Liu, J.-N.; Chi, P.; Lai, W.; Wu, J.-S.; Guo, Q.-L.; Lin, D.-D.; Liu, Y.; et al. Liver transplantation in acute-on-chronic liver failure patients with high model for end-stage liver disease (MELD) scores-a single center experience of 100 consecutive cases. *J. Surg. Res.* **2013**, *18*, 936–943. [CrossRef]
11. Kidambi, S.; Yarmush, R.S.; Novik, E.; Chao, P.; Yarmush, M.L.; Nahmias, Y. Oxygen-mediated enhancement of primary hepatocyte metabolism, functional polarization, gene expression, and drug clearance. *Proc. Natl. Acad Sci. USA* **2009**, *106*, 15714–15719. [CrossRef] [PubMed]
12. Habibullah, C.M.; Syed, I.H.; Qamar, A.; Taher-Uz, Z. Human fetal hepatocyte transplantation in patients with fulminant hepatic failure. *Transplantation* **1994**, *58*, 951–952. [CrossRef] [PubMed]
13. Kobayashi, N.; Noguchi, H.; Matsumura, T.; Watanabe, T.; Totsugawa, T.; Fujiwara, T.; Tanaka, N. Establishment of a tightly regulated immortalized human hepatocyte cell line for the development of bioartificial liver. *Hepatology* **2000**, *32*, 611a.
14. Mito, M.; Kusano, M. Hepatocyte Transplantation in Man. *Cell Transpl.* **1993**, *2*, 65–74. [CrossRef]
15. Strom, S.C.; Fisher, R.A.; Thompson, M.T.; Sanyal, A.J.; Cole, P.E.; Ham, J.M.; Posner, M.P. Hepatocyte transplantation as a bridge to orthotopic liver transplantation in terminal liver failure. *Transplantation* **1997**, *63*, 559–569. [CrossRef]
16. Kisseleva, T.; Gigante, E.; Brenner, D.A. Recent advances in liver stem cell therapy. *Curr. Opin. Gastroenterol.* **2010**, *26*, 395–402. [CrossRef]
17. Zhang, W.; Tucker-Kellogg, L.; Narmada, B.C.; Venkatraman, L.; Chang, S.; Lu, Y.; Tan, N.; White, J.K.; Jia, R.; Bhowmick, S.S.; et al. Cell-delivery therapeutics for liver regeneration. *Adv. Drug Deliv. Rev.* **2010**, *62*, 814–826. [CrossRef]
18. Khan, A.A.; Shaik, M.V.; Parveen, N.; Rajendraprasad, A.; Aleem, M.A.; Habeeb, M.A.; Srinivas, G.; Raj, T.A.; Tiwari, S.K.; Kumaresan, K.; et al. Human Fetal Liver-Derived Stem Cell Transplantation as Supportive Modality in the Management of End-Stage Decompensated Liver Cirrhosis. *Cell Transpl.* **2010**, *19*, 409–418. [CrossRef]

19. Cai, J.; Zhao, Y.; Liu, Y.; Ye, F.; Song, Z.; Qin, H.; Meng, S.; Chen, Y.; Zhou, R.; Song, X.; et al. Directed differentiation of human embryonic stem cells into functional hepatic cells. *Hepatology* **2007**, *45*, 1229–1239. [CrossRef]

20. Huch, M.; Dorrell, C.; Boj, S.F.; van Es, J.H.; Li, V.S.; van de Wetering, M.; Sato, T.; Hamer, K.; Sasaki, N.; Finegold, M.J.; et al. In vitro expansion of single Lgr5+ liver stem cells induced by Wnt-driven regeneration. *Nature* **2013**, *494*, 247–250. [CrossRef]

21. Si-Tayeb, K.; Noto, F.K.; Nagaoka, M.; Li, J.; Battle, M.A.; Duris, C.; North, P.E.; Dalton, S.; Duncan, S.A. Highly efficient generation of human hepatocyte-like cells from induced pluripotent stem cells. *Hepatology* **2010**, *51*, 297–305. [CrossRef] [PubMed]

22. Stock, P.; Bruckner, S.; Ebensing, S.; Hempel, M.; Dollinger, M.M.; Christ, B. The generation of hepatocytes from mesenchymal stem cells and engraftment into murine liver. *Nat. Protoc.* **2010**, *5*, 617–627. [CrossRef] [PubMed]

23. Huch, M.; Gehart, H.; van Boxtel, R.; Hamer, K.; Blokzijl, F.; Verstegen, M.M.; Ellis, E.; van Wenum, M.; Fuchs, S.A.; de Ligt, J.; et al. Long-term culture of genome-stable bipotent stem cells from adult human liver. *Cell* **2015**, *160*, 299–312. [CrossRef] [PubMed]

24. Chen, C.; Soto-Gutierrez, A.; Baptista, P.M.; Spee, B. Biotechnology Challenges to In Vitro Maturation of Hepatic Stem Cells. *Gastroenterology* **2018**, *154*, 1258–1272. [CrossRef] [PubMed]

25. Levy, G.; Bomze, D.; Heinz, S.; Ramachandran, S.D.; Noerenberg, A.; Cohen, M.; Shibolet, O.; Sklan, E.; Braspenning, J.; Nahmias, Y. Long-term culture and expansion of primary human hepatocytes. *Nat. Biotechnol.* **2015**, *33*, 1264–1271. [CrossRef] [PubMed]

26. Zhang, K.; Zhang, L.; Liu, W.; Ma, X.; Cen, J.; Sun, Z.; Wang, C.; Feng, S.; Zhang, Z.; Yue, L.; et al. In Vitro Expansion of Primary Human Hepatocytes with Efficient Liver Repopulation Capacity. *Cell Stem Cell* **2018**. [CrossRef]

27. Hu, H.; Gehart, H.; Artegiani, B.; LÖpez-Iglesias, C.; Dekkers, F.; Basak, O.; van Es, J.; Chuva de Sousa Lopes, S.M.; Begthel, H.; Korving, J.; et al. Long-Term Expansion of Functional Mouse and Human Hepatocytes as 3D Organoids. *Cell* **2018**, *175*, 1591–1606. [CrossRef] [PubMed]

28. Sivashanmugam, A.; Arun Kumar, R.; Vishnu Priya, M.; Nair, S.V.; Jayakumar, R. An overview of injectable polymeric hydrogels for tissue engineering. *Eur. Polym. J.* **2015**, *72*, 543–565. [CrossRef]

29. Bedian, L.; Villalba-Rodriguez, A.M.; Hernandez-Vargas, G.; Parra-Saldivar, R.; Iqbal, H.M. Bio-based materials with novel characteristics for tissue engineering applications-A review. *Int. J. Biol. Macromol.* **2017**, *98*, 837–846. [CrossRef]

30. Bahram, M.; Mohseni, N.; Moghtader, M. An Introduction to Hydrogels and Some Recent Applications. In *Emerging Concepts in Analysis and Applications of Hydrogels*; InTech: London, UK, 2016; ISBN 978-953-51-2510-5.

31. Lee, K.Y.; Mooney, D.J. Hydrogels for Tissue Engineering. *Chem. Rev.* **2001**, *101*. [CrossRef]

32. Curvello, R.; Raghuwanshi, V.S.; Garnier, G. Engineering nanocellulose hydrogels for biomedical applications. *Adv. Colloid Interface Sci.* **2019**, *267*, 47–61. [CrossRef] [PubMed]

33. Chiti, M.C.; Dolmans, M.M.; Donnez, J.; Amorim, C.A. Fibrin in Reproductive Tissue Engineering: A Review on Its Application as a Biomaterial for Fertility Preservation. *Ann. Biomed Eng.* **2017**, *45*, 1650–1663. [CrossRef] [PubMed]

34. Dong, C.; Lv, Y. Application of Collagen Scaffold in Tissue Engineering: Recent Advances and New Perspectives. *Polymers* **2016**, *8*, 42. [CrossRef] [PubMed]

35. Dash, T.K.; Konkimalla, V.B. Poly-small je, Ukrainian-caprolactone based formulations for drug delivery and tissue engineering: A review. *J. Control. Release* **2012**, *158*, 15–33. [CrossRef] [PubMed]

36. Kumar, A.; Han, S.S. PVA-based hydrogels for tissue engineering A review. *Int. J. Polym. Mater. Polym. Biomater.* **2017**, *66*, 159–182. [CrossRef]

37. Annabi, N.; Tamayol, A.; Uquillas, J.A.; Akbari, M.; Bertassoni, L.E.; Cha, C.; Camci-Unal, G.; Dokmeci, M.R.; Peppas, N.A.; Khademhosseini, A. 25th Anniversary Article: Rational Design and Applications of Hydrogels in Regenerative Medicine. *Adv. Mater.* **2014**, *26*, 85–124. [CrossRef] [PubMed]

38. Drury, J.L.; Mooney, D.J. Hydrogels for tissue engineering: Scaffold design variables and applications. *Biomaterials* **2003**, *24*, 4337–4351. [CrossRef]

39. Khorshidi, S.; Karkhaneh, A. A review on gradient hydrogel/fiber scaffolds for osteochondral regeneration. *J. Tissue Eng. Regen. Med.* **2018**, *12*, e1974–e1990. [CrossRef] [PubMed]

40. Gradinaru, V.; Treweek, J.; Overton, K.; Deisseroth, K. Hydrogel-Tissue Chemistry: Principles and Applications. *Annu. Rev. Biophys.* **2018**, *47*, 355–376. [CrossRef] [PubMed]

41. Jung, J. Initiation of Mammalian Liver Development from Endoderm by Fibroblast Growth Factors. *Science* **1999**, *284*, 1998–2003. [CrossRef]

42. Bohm, F.; Kohler, U.A.; Speicher, T.; Werner, S. Regulation of liver regeneration by growth factors and cytokines. *EMBO Mol. Med.* **2010**, *2*, 294–305. [CrossRef] [PubMed]

43. Bosch, F.X.; Ribes, J.; Díaz, M.; Cléries, R. Primary liver cancer: Worldwide incidence and trends. *Gastroenterology* **2004**, *127*, S5–S16. [CrossRef] [PubMed]

44. Rebecca, T. Liver regeneration: from myth to mechanism. *Nat. Rev. Mol. Cell Biol.* **2004**, *5*, 836–847.

45. Court, F.G.; Wemyss-Holden, S.A.; Dennison, A.R.; Maddern, G.J. The mystery of liver regeneration. *Br. J. Surg.* **2002**, *89*, 1089–1095. [CrossRef] [PubMed]

46. Yokoyama, Y.; Nagino, M.; Nimura, Y. Mechanisms of Hepatic Regeneration Following Portal Vein Embolization and Partial Hepatectomy: A Review. *World J. Surg.* **2007**, *31*, 367–374. [CrossRef] [PubMed]

47. Fausto, N.; Riehle, K.J. Mechanisms of liver regeneration and their clinical implications. *J. Hepatobiliary Pancreat. Surg.* **2005**, *12*, 181–189. [CrossRef] [PubMed]

48. Fausto, N.; Campbell, J.S.; Riehle, K.J. Liver regeneration. *Hepatology* **2006**, *43*, S45–S53. [CrossRef] [PubMed]

49. Michalopoulos, G.K. Liver regeneration. *J. Cell Physiol.* **2007**, *213*, 286–300. [CrossRef]

50. Duncan, A.W.; Dorrell, C.; Grompe, M. Stem cells and liver regeneration. *Gastroenterology* **2009**, *137*, 466–481. [CrossRef]

51. Riehle, K.J.; Dan, Y.Y.; Campbell, J.S.; Fausto, N. New concepts in liver regeneration. *J. Gastroenterol. Hepatol.* **2011**, *26* (Suppl. 1), 203–212. [CrossRef]

52. DeLeve, L.D. Liver sinusoidal endothelial cells and liver regeneration. *J. Clin. Investig.* **2013**, *123*, 1861–1866. [CrossRef] [PubMed]

53. Forbes, S.J.; Newsome, P.N. Liver regeneration-mechanisms and models to clinical application. *Nat. Rev. Gastroenterol Hepatol* **2016**, *13*, 473–485. [CrossRef] [PubMed]

54. Timchenko, N.A. Aging and liver regeneration. *Trends Endocrinol. Metab.* **2009**, *20*, 171–176. [CrossRef] [PubMed]

55. Zhou, Q.; Li, L.; Li, J. Stem cells with decellularized liver scaffolds in liver regeneration and their potential clinical applications. *Liver Int.* **2015**, *35*, 687–694. [CrossRef] [PubMed]

56. Souza, F.S.D.; Botelho, M.C.S.N.; Lisbôa, R.S. Hepatology: Hepatology Principles and Practice: History, Morphology, Biochemistry, Diagnostics, Clinic, Therapy. *J. Am. Med Assoc.* **2002**, *288*, 1–15.

57. Willemse, J.; Lieshout, R.; van der Laan, L.J.W.; Verstegen, M.M.A. From organoids to organs: Bioengineering liver grafts from hepatic stem cells and matrix. *Best Pract. Res. Clin. Gastroenterol.* **2017**, *31*, 151–159. [CrossRef] [PubMed]

58. Gordillo, M.; Evans, T.; Gouon-Evans, V. Orchestrating liver development. *Development* **2015**, *142*, 2094–2108. [CrossRef]

59. Racanelli, V.; Rehermann, B. The liver as an immunological organ. *Hepatology* **2006**, *43*, S54–S62. [CrossRef]

60. Lin, X.; Horng, M.; Sun, Y.; Shiesh, S.C.; Chow, N.H.; Guo, X.Z. Computer morphometry for quantitative measurement of liver fibrosis: Comparison with Knodell's score, colorimetry and conventional description reports. *J. Gastroenterol. Hepatol.* **2010**, *13*, 75–80. [CrossRef]

61. Martinez-Hernandez, A.; Amenta, P.S. The hepatic extracellular matrix. *Virchows Arch. A* **1993**, *423*, 77–84. [CrossRef]

62. Thiery, J.P.; Duband, J.L.; Tucker, G.C. Cell Migration in the Vertebrate Embryo: Role of Cell Adhesion and Tissue Environment in Pattern Formation. *Annu. Rev. Cell Biol.* **1985**, *1*, 91–113. [CrossRef] [PubMed]

63. Kleinman, H.K.; Philp, D.; Hoffman, M.P. Role of the extracellular matrix in morphogenesis. *Curr. Opin. Biotechnol.* **2003**, *14*, 526–532. [CrossRef] [PubMed]

64. Bedossa, P.; Paradis, V. Liver extracellular matrix in health and disease. *J. Pathol.* **2003**, *200*, 504–515. [CrossRef] [PubMed]

65. Hynes, R.O. The extracellular matrix: Not just pretty fibrils. *Science* **2009**, *326*, 1216–1219. [CrossRef] [PubMed]

66. Baiocchini, A.; Montaldo, C.; Conigliaro, A.; Grimaldi, A.; Correani, V.; Mura, F.; Ciccosanti, F.; Rotiroti, N.; Brenna, A.; Montalbano, M.; et al. Extracellular Matrix Molecular Remodeling in Human Liver Fibrosis Evolution. *PLoS ONE* **2016**, *11*, e0151736. [CrossRef]

67. Martinez-Hernandez, A. The hepatic extracellular matrix. I. Electron immunohistochemical studies in normal rat liver. *Lab. Investig.* **1984**, *51*, 57–74.
68. Rojkind, M.; Ponce-Noyola, P. The Extracellular Matrix of the Liver. *Collagen Relat. Res.* **1982**, *2*, 151–175. [CrossRef]
69. Marrone, G.; Shah, V.H.; Gracia-Sancho, J. Sinusoidal communication in liver fibrosis and regeneration. *J. Hepatol.* **2016**, *65*, 608–617. [CrossRef]
70. Zhu, C.; Coombe, D.R.; Zheng, M.H.; Yeoh, G.C.; Li, L. Liver progenitor cell interactions with the extracellular matrix. *J. Tissue Eng. Regen. Med.* **2013**, *7*, 757–766. [CrossRef]
71. Klaas, M.; Kangur, T.; Viil, J.; Maemets-Allas, K.; Minajeva, A.; Vadi, K.; Antsov, M.; Lapidus, N.; Jarvekulg, M.; Jaks, V. The alterations in the extracellular matrix composition guide the repair of damaged liver tissue. *Sci. Rep.* **2016**, *6*, 27398. [CrossRef]
72. Bemmelen, J.M.V. Der Hydrogel und das kristallinische Hydrat des Kupferoxydes. *Z. Anorg. Chem.* **1894**, *5*, 466. [CrossRef]
73. Danno, A. Gel Formation of Aqueous Solution of Polyvinyl Alcohol Irradiated by Gamma Rays from Cobalt-60. *J. Phys. Soc. Jpn.* **1958**, *13*, 722–727. [CrossRef]
74. Wichterle, O.; Lim, D. Hydrophilic Gels for Biological Use. *Nature* **1960**, *185*, 117–118. [CrossRef]
75. Buwalda, S.J.; Boere, K.W.; Dijkstra, P.J.; Feijen, J.; Vermonden, T.; Hennink, W.E. Hydrogels in a historical perspective: From simple networks to smart materials. *J. Control. Release* **2014**, *190*, 254–273. [CrossRef] [PubMed]
76. Vishwakarma, S.K.; Bardia, A.; Lakkireddy, C.; Raju, N.; Paspala, S.A.B.; Habeeb, M.A.; Khan, A.A. Intraperitoneal transplantation of bioengineered humanized liver grafts supports failing liver in acute condition. *Mater. Sci. Eng. C Mater Biol. Appl.* **2019**, *98*, 861–873. [CrossRef] [PubMed]
77. Vasanthan, K.S.; Subramanian, A.; Krishnan, U.M.; Sethuraman, S. Role of biomaterials, therapeutic molecules and cells for hepatic tissue engineering. *Biotechnol. Adv.* **2012**, *30*, 742–752. [CrossRef]
78. Barbucci, R. *Hydrogels: Biological Properties and Applications*; Springer: Milan, Italy, 2009; pp. 1–179.
79. Wang, X.H.; Li, D.P.; Wang, W.J.; Feng, Q.L.; Cui, F.Z.; Xu, Y.X.; Song, X.H.; van der Werf, M. Crosslinked collagen/chitosan matrix for artificial livers. *Biomaterials* **2003**, *24*, 3213–3220. [CrossRef]
80. Yu, X.; Bichtelen, A.; Wang, X.; Yan, Y.; Lin, F.; Xiong, Z.; Wu, R.; Zhang, R.; Lu, Q. Collagen/Chitosan/Heparin Complex with Improved Biocompatibility for Hepatic Tissue Engineering. *J. Bioact. Compat. Polym.* **2016**, *20*, 15–28. [CrossRef]
81. Nair, L.S.; Laurencin, C.T. Biodegradable polymers as biomaterials. *Prog. Polym. Sci.* **2007**, *32*, 762–798. [CrossRef]
82. Bhattacharya, M.; Malinen, M.M.; Lauren, P.; Lou, Y.R.; Kuisma, S.W.; Kanninen, L.; Lille, M.; Corlu, A.; GuGuen-Guillouzo, C.; Ikkala, O.; et al. Nanofibrillar cellulose hydrogel promotes three-dimensional liver cell culture. *J. Control. Release* **2012**, *164*, 291–298. [CrossRef]
83. Park, U.J.; Jeong, W.; Kwon, S.Y.; Kim, Y.; Choi, K.; Kim, H.T.; Son, D. Fabrication of a Novel Absorbable Vascular Anastomosis Device and Testing in a Pig Liver Transplantation Model. *Ann. Biomed. Eng.* **2019**, *47*, 1063–1077. [CrossRef] [PubMed]
84. Li, J.; Tao, R.; Wu, W.; Cao, H.; Xin, J.; Li, J.; Guo, J.; Jiang, L.; Gao, C.; Demetriou, A.A.; et al. 3D PLGA scaffolds improve differentiation and function of bone marrow mesenchymal stem cell-derived hepatocytes. *Stem Cells Dev.* **2010**, *19*, 1427–1436. [CrossRef] [PubMed]
85. Brandl, F.; Sommer, F.; Goepferich, A. Rational design of hydrogels for tissue engineering: Impact of physical factors on cell behavior. *Biomaterials* **2007**, *28*, 134–146. [CrossRef] [PubMed]
86. Bomo, J.; Ezan, F.; Tiaho, F.; Bellamri, M.; Langouet, S.; Theret, N.; Baffet, G. Increasing 3D Matrix Rigidity Strengthens Proliferation and Spheroid Development of Human Liver Cells in a Constant Growth Factor Environment. *J. Cell Biochem.* **2016**, *117*, 708–720. [CrossRef] [PubMed]
87. Desai, S.S.; Tung, J.C.; Zhou, V.X.; Grenert, J.P.; Malato, Y.; Rezvani, M.; Espanol-Suner, R.; Willenbring, H.; Weaver, V.M.; Chang, T.T. Physiological ranges of matrix rigidity modulate primary mouse hepatocyte function in part through hepatocyte nuclear factor 4 alpha. *Hepatology* **2016**, *64*, 261–275. [CrossRef] [PubMed]
88. Wells, R.G. The role of matrix stiffness in regulating cell behavior. *Hepatology* **2008**, *47*, 1394–1400. [CrossRef] [PubMed]

89. Gong, H.; Agustin, J.; Wootton, D.; Zhou, J.G. Biomimetic design and fabrication of porous chitosan-gelatin liver scaffolds with hierarchical channel network. *J. Mater. Sci. Mater. Med.* **2014**, *25*, 113–120. [CrossRef] [PubMed]

90. Capone, S.H.; Dufresne, M.; Rechel, M.; Fleury, M.J.; Salsac, A.V.; Paullier, P.; Daujat-Chavanieu, M.; Legallais, C. Impact of alginate composition: From bead mechanical properties to encapsulated HepG2/C3A cell activities for in vivo implantation. *PLoS ONE* **2013**, *8*, e62032. [CrossRef] [PubMed]

91. Annabi, N.; Nichol, J.W.; Zhong, X.; Ji, C.; Koshy, S.; Khademhosseini, A.; Dehghani, F. Controlling the Porosity and Microarchitecture of Hydrogels for Tissue Engineering. *Tissue Eng. Part B Rev.* **2010**, *16*, 371–383. [CrossRef]

92. Zhao, H.; Xu, K.; Zhu, P.; Wang, C.; Chi, Q. Smart hydrogels with high tunability of stiffness as a biomimetic cell carrier. *Cell Biol. Int.* **2019**, *43*, 84–97. [CrossRef]

93. Park, T.G. Perfusion culture of hepatocytes within galactose-derivatized biodegradable poly(lactide-co-glycolide) scaffolds prepared by gas foaming of effervescent salts. *J. Biomed. Mater. Res. Part B Appl. Biomater.* **2010**, *59*, 127–135. [CrossRef] [PubMed]

94. Park, K.H.; Bae, Y.H. Phenotype of Hepatocyte Spheroids in Arg-Gly-Asp (RGD) Containing a Thermo-Reversible Extracellular Matrix. *Biosci. Biotechnol. Biochem.* **2002**, *66*, 1473–1478. [CrossRef] [PubMed]

95. Ma, C.Y.J.; Kumar, R.; Xu, X.Y.; Mantalaris, A. A combined fluid dynamics, mass transport and cell growth model for a three-dimensional perfused biorector for tissue engineering of haematopoietic cells. *Biochem. Eng. J.* **2007**, *35*, 1–11. [CrossRef]

96. Wu, C.; Pan, J.; Bao, Z.; Yu, Y. Fabrication and characterization of chitosan microcarrier for hepatocyte culture. *J. Mater. Sci. Mater. Med.* **2007**, *18*, 2211–2214. [CrossRef] [PubMed]

97. Tripathi, A.; Melo, J.S. Preparation of a sponge-like biocomposite agarose–chitosan scaffold with primary hepatocytes for establishing an in vitro 3D liver tissue model. *RSC Adv.* **2015**, *5*, 30701–30710. [CrossRef]

98. Jaspers, M.; Dennison, M.; Mabesoone, M.F.; MacKintosh, F.C.; Rowan, A.E.; Kouwer, P.H. Ultra-responsive soft matter from strain-stiffening hydrogels. *Nat. Commun.* **2014**, *5*, 5808. [CrossRef] [PubMed]

99. Necas, J.; Brauner, P.; Kolar, J.; Bartosikova, L. Hyaluronic acid (hyaluronan)—A review. *Vet. Med.* **2008**, *53*, 397–411. [CrossRef]

100. Allison, D.D.; Grande-Allen, K.J. Hyaluronan: A powerful tissue engineering tool. *Tissue Eng.* **2006**, *12*, 2131–2140. [CrossRef]

101. Augst, A.D.; Kong, H.J.; Mooney, D.J. Alginate hydrogels as biomaterials. *Macromol. Biosci.* **2006**, *6*, 623–633. [CrossRef]

102. Lee, K.Y.; Mooney, D.J. Alginate: Properties and biomedical applications. *Prog. Polym. Sci.* **2012**, *37*, 106–126. [CrossRef]

103. Liu, C.Z.; Xia, Z.D.; Han, Z.W.; Hulley, P.A.; Triffitt, J.T.; Czernuszka, J.T. Novel 3D collagen scaffolds fabricated by indirect printing technique for tissue engineering. *J. Biomed. Mater. Res. B Appl. Biomater.* **2008**, *85*, 519–528. [CrossRef] [PubMed]

104. Van Vlierberghe, S.; Dubruel, P.; Schacht, E. Biopolymer-based hydrogels as scaffolds for tissue engineering applications: A review. *Biomacromolecules* **2011**, *12*, 1387–1408. [CrossRef]

105. Klotz, B.J.; Gawlitta, D.; Rosenberg, A.; Malda, J.; Melchels, F.P.W. Gelatin-Methacryloyl Hydrogels: Towards Biofabrication-Based Tissue Repair. *Trends Biotechnol.* **2016**, *34*, 394–407. [CrossRef] [PubMed]

106. Kleinman, H.K.; Martin, G.R. Matrigel: Basement membrane matrix with biological activity. *Semin. Cancer Biol.* **2005**, *15*, 378–386. [CrossRef] [PubMed]

107. Snyder, J.E.; Hamid, Q.; Wang, C.; Chang, R.; Emami, K.; Wu, H.; Sun, W. Bioprinting cell-laden matrigel for radioprotection study of liver by pro-drug conversion in a dual-tissue microfluidic chip. *Biofabrication* **2011**, *3*, 034112. [CrossRef]

108. Hughes, C.S.; Postovit, L.M.; Lajoie, G.A. Matrigel: A complex protein mixture required for optimal growth of cell culture. *Proteomics* **2010**, *10*, 1886–1890. [CrossRef]

109. Currie, L.J.; Sharpe, J.R.; Martin, R. The use of fibrin glue in skin grafts and tissue-engineered skin replacements: A review. *Plast. Reconstr. Surg.* **2001**, *108*, 1713–1726. [CrossRef]

110. Ahmed, T.A.; Griffith, M.; Hincke, M. Characterization and inhibition of fibrin hydrogel-degrading enzymes during development of tissue engineering scaffolds. *Tissue Eng.* **2007**, *13*, 1469–1477. [CrossRef]

111. Kalia, V.C.; Ray, S.; Patel, S.K.S.; Singh, M.; Singh, G.P. The Dawn of Novel Biotechnological Applications of Polyhydroxyalkanoates. In *Biotechnological Applications of Polyhydroxyalkanoates*; Kalia, V., Ed.; Springer: Singapore, 2019.

112. Zhang, Y.S.; Yue, K.; Aleman, J.; Moghaddam, K.M.; Bakht, S.M.; Yang, J.; Jia, W.; Dell'Erba, V.; Assawes, P.; Shin, S.R.; et al. 3D Bioprinting for Tissue and Organ Fabrication. *Ann. Biomed. Eng.* **2017**, *45*, 148–163. [CrossRef]

113. Mazzocchi, A.; Devarasetty, M.; Huntwork, R.; Soker, S.; Skardal, A. Optimization of collagen type I-hyaluronan hybrid bioink for 3D bioprinted liver microenvironments. *Biofabrication* **2018**, *11*, 015003. [CrossRef]

114. Wang, X.; Yan, Y.; Xiong, Z.; Lin, F.; Wu, R.; Zhang, R.; Lu, Q. Preparation and evaluation of ammonia-treated collagen/chitosan matrices for liver tissue engineering. *J. Biomed. Mater. Res. B Appl. Biomater.* **2005**, *75*, 91–98. [CrossRef] [PubMed]

115. Wang, X.; Yan, Y.; Lin, F.; Xiong, Z.; Wu, R.; Zhang, R.; Lu, Q. Preparation and characterization of a collagen/chitosan/heparin matrix for an implantable bioartificial liver. *J. Biomater. Sci. Polym. Ed.* **2012**, *16*, 1063–1080. [CrossRef]

116. Wang, X.; Yan, Y.; Pan, Y.; Xiong, Z.; Liu, H.; Cheng, J.; Liu, F.; Lin, F.; Wu, R.; Zhang, R.; et al. Generation of Three-Dimensional Hepatocyte-Gelatin Structures with Rapid Prototyping System. *Tissue Eng.* **2006**, *12*, 83–90. [CrossRef] [PubMed]

117. Hussein, K.H.; Park, K.M.; Kang, K.S.; Woo, H.M. Heparin-gelatin mixture improves vascular reconstruction efficiency and hepatic function in bioengineered livers. *Acta Biomater.* **2016**, *38*, 82–93. [CrossRef] [PubMed]

118. Wei, X.; Xiaohong, W.; Yongnian, Y.; Renji, Z. A Polyurethane-Gelatin Hybrid Construct for Manufacturing Implantable Bioartificial Livers. *J. Bioact. Compat. Polym.* **2008**, *23*, 409–422. [CrossRef]

119. Rozario, T.; DeSimone, D.W. The extracellular matrix in development and morphogenesis: A dynamic view. *Dev. Biol.* **2010**, *341*, 126–140. [CrossRef] [PubMed]

120. Suri, S.; Han, L.H.; Zhang, W.; Singh, A.; Chen, S.; Schmidt, C.E. Solid freeform fabrication of designer scaffolds of hyaluronic acid for nerve tissue engineering. *Biomed. Microdevices* **2011**, *13*, 983–993. [CrossRef] [PubMed]

121. Turner, W.S.; Schmelzer, E.; McClelland, R.; Wauthier, E.; Chen, W.; Reid, L.M. Human hepatoblast phenotype maintained by hyaluronan hydrogels. *J. Biomed. Mater. Res. B Appl. Biomater.* **2007**, *82*, 156–168. [CrossRef]

122. Christoffersson, J.; Aronsson, C.; Jury, M.; Selegard, R.; Aili, D.; Mandenius, C.F. Fabrication of modular hyaluronan-PEG hydrogels to support 3D cultures of hepatocytes in a perfused liver-on-a-chip device. *Biofabrication* **2018**. [CrossRef]

123. Bruns, H.; Kneser, U.; Holzhüter, S.; Roth, B.; Kluth, J.; Kaufmann, P.M.; Kluth, D.; Fiegel, H.C. Injectable Liver A Novel Approach Using Fibrin Gel as a Matrix for Culture and Intrahepatic Transplantation of Hepatocytes. *Tissue Eng.* **2005**, *11*, 1718–1726. [CrossRef]

124. Wang, X.; Liu, C. Fibrin Hydrogels for Endothelialized Liver Tissue Engineering with a Predesigned Vascular Network. *Polymers* **2018**, *10*, 48. [CrossRef] [PubMed]

125. Hickey, R.D.; Naugler, W.E. Ectopic expansion of engineered human liver tissue seeds using mature cell populations. *Hepatology* **2018**, *67*, 2465–2467. [CrossRef]

126. Jammalamadaka, U.; Tappa, K. Recent Advances in Biomaterials for 3D Printing and Tissue Engineering. *J. Funct. Biomater.* **2018**, *9*, 22. [CrossRef]

127. Kim, Y.; Kang, K.; Jeong, J.; Paik, S.S.; Kim, J.S.; Park, S.A.; Kim, W.D.; Park, J.; Choi, D. Three-dimensional (3D) printing of mouse primary hepatocytes to generate 3D hepatic structure. *Ann. Surg. Treat. Res.* **2017**, *92*, 67–72. [CrossRef] [PubMed]

128. Tong, X.F.; Zhao, F.Q.; Ren, Y.Z.; Zhang, Y.; Cui, Y.L.; Wang, Q.S. Injectable hydrogels based on glycyrrhizin, alginate, and calcium for three-dimensional cell culture in liver tissue engineering. *J. Biomed. Mater. Res. A* **2018**, *106*, 3292–3302. [CrossRef] [PubMed]

129. Lee, K.H.; Shin, S.J.; Kim, C.B.; Kim, J.K.; Cho, Y.W.; Chung, B.G.; Lee, S.H. Microfluidic synthesis of pure chitosan microfibers for bio-artificial liver chip. *Lab. Chip.* **2010**, *10*, 1328–1334. [CrossRef] [PubMed]

130. Jiankang, H.; Dichen, L.; Yaxiong, L.; Bo, Y.; Hanxiang, Z.; Qin, L.; Bingheng, L.; Yi, L. Preparation of chitosan-gelatin hybrid scaffolds with well-organized microstructures for hepatic tissue engineering. *Acta Biomater.* **2009**, *5*, 453–461. [CrossRef] [PubMed]

131. Rajendran, D.; Hussain, A.; Yip, D.; Parekh, A.; Shrirao, A.; Cho, C.H. Long-term liver-specific functions of hepatocytes in electrospun chitosan nanofiber scaffolds coated with fibronectin. *J. Biomed. Mater. Res. A* **2017**, *105*, 2119–2128. [CrossRef]

132. Su, Z.; Li, P.; Wu, B.; Ma, H.; Wang, Y.; Liu, G.; Zeng, H.; Li, Z.; Wei, X. PHBVHHx scaffolds loaded with umbilical cord-derived mesenchymal stem cells or hepatocyte-like cells differentiated from these cells for liver tissue engineering. *Mater. Sci. Eng. C Mater. Biol. Appl.* **2014**, *45*, 374–382. [CrossRef]

133. Li, P.; Zhang, J.; Liu, J.; Ma, H.; Liu, J.; Lie, P.; Wang, Y.; Liu, G.; Zeng, H.; Li, Z.; et al. Promoting the recovery of injured liver with poly (3-hydroxybutyrate-co-3-hydroxyvalerate-co-3-hydroxyhexanoate) scaffolds loaded with umbilical cord-derived mesenchymal stem cells. *Tissue Eng. Part A* **2015**, *21*, 603–615. [CrossRef] [PubMed]

134. Markstedt, K.; Mantas, A.; Tournier, I.; Martinez Avila, H.; Hagg, D.; Gatenholm, P. 3D Bioprinting Human Chondrocytes with Nanocellulose-Alginate Bioink for Cartilage Tissue Engineering Applications. *Biomacromolecules* **2015**, *16*, 1489–1496. [CrossRef] [PubMed]

135. Wu, Y.; Lin, Z.Y.; Wenger, A.C.; Tam, K.C.; Tang, X. 3D bioprinting of liver-mimetic construct with alginate/cellulose nanocrystal hybrid bioink. *Bioprinting* **2018**, *9*, 1–6. [CrossRef]

136. Malinen, M.M.; Kanninen, L.K.; Corlu, A.; Isoniemi, H.M.; Lou, Y.R.; Yliperttula, M.L.; Urtti, A.O. Differentiation of liver progenitor cell line to functional organotypic cultures in 3D nanofibrillar cellulose and hyaluronan-gelatin hydrogels. *Biomaterials* **2014**, *35*, 5110–5121. [CrossRef] [PubMed]

137. Jiang, S.; Liu, S.; Feng, W. PVA hydrogel properties for biomedical application. *J. Mech. Behav. Biomed. Mater.* **2011**, *4*, 1228–1233. [CrossRef] [PubMed]

138. Kouwer, P.H.; Koepf, M.; Le Sage, V.A.; Jaspers, M.; van Buul, A.M.; Eksteen-Akeroyd, Z.H.; Woltinge, T.; Schwartz, E.; Kitto, H.J.; Hoogenboom, R.; et al. Responsive biomimetic networks from polyisocyanopeptide hydrogels. *Nature* **2013**, *493*, 651–655. [CrossRef] [PubMed]

139. Hammink, R.; Eggermont, L.J.; Zisis, T.; Tel, J.; Figdor, C.G.; Rowan, A.E.; Blank, K.G. Affinity-Based Purification of Polyisocyanopeptide Bioconjugates. *Bioconjug. Chem.* **2017**, *28*, 2560–2568. [CrossRef] [PubMed]

140. Utech, S.; Boccaccini, A.R. A review of hydrogel-based composites for biomedical applications: enhancement of hydrogel properties by addition of rigid inorganic fillers. *J. Mater. Sci.* **2016**, *51*, 271–310. [CrossRef]

141. Saito, E.; Kang, H.; Taboas, J.M.; Diggs, A.; Flanagan, C.L.; Hollister, S.J. Experimental and computational characterization of designed and fabricated 50:50 PLGA porous scaffolds for human trabecular bone applications. *J. Mater. Sci. Mater. Med.* **2010**, *21*, 2371–2383. [CrossRef] [PubMed]

142. Sheridan, M.H.; Shea, L.D.; Peters, M.C.; Mooney, D.J. Bioabsorbable polymer scaffolds for tissue engineering capable. *J. Control. Release* **2000**, *64*, 91–102. [CrossRef]

143. Middleton, J.C.; Tipton, A.J. Synthetic biodegradable polymers as orthopedic devices. *Biomaterials* **2000**, *21*, 2335–2346. [CrossRef]

144. Lv, Q.; Hu, K.; Feng, Q.; Cui, F.; Cao, C. Preparation and characterization of PLA/fibroin composite and culture of HepG2 (human hepatocellular liver carcinoma cell line) cells. *Compos. Sci. Technol.* **2007**, *67*, 3023–3030. [CrossRef]

145. Siddiqui, N.; Asawa, S.; Birru, B.; Baadhe, R.; Rao, S. PCL-Based Composite Scaffold Matrices for Tissue Engineering Applications. *Mol. Biotechnol.* **2018**, *60*, 506–532. [CrossRef] [PubMed]

146. JANG, T.-S.; Jung, H.-D.; Pan, H.M.; Han, W.T. 3D printing of hydrogel composite systems: Recent advances in technology for tissue engineering. *Int. J. Bioprinting* **2018**. [CrossRef]

147. Yoon No, D.; Lee, K.H.; Lee, J.; Lee, S.H. 3D liver models on a microplatform: Well-defined culture, engineering of liver tissue and liver-on-a-chip. *Lab. Chip.* **2015**, *15*, 3822–3837. [CrossRef] [PubMed]

148. Perez, R.A.; Jung, C.R.; Kim, H.W. Biomaterials and Culture Technologies for Regenerative Therapy of Liver Tissue. *Adv. Healthc. Mater.* **2017**, *6*. [CrossRef]

149. Schepers, A.; Li, C.; Chhabra, A.; Seney, B.T.; Bhatia, S. Engineering a perfusable 3D human liver platform from iPS cells. *Lab. Chip.* **2016**, *16*, 2644–2653. [CrossRef] [PubMed]

150. Ng, S.S.; Xiong, A.; Nguyen, K.; Masek, M.; No, D.Y.; Elazar, M.; Shteyer, E.; Winters, M.A.; Voedisch, A.; Shaw, K.; et al. Long-term culture of human liver tissue with advanced hepatic functions. *JCI Insight.* **2017**, *2*. [CrossRef]

151. Celikkin, N.; Simó Padial, J.; Costantini, M.; Hendrikse, H.; Cohn, R.; Wilson, C.; Rowan, A.; Święszkowski, W. 3D Printing of Thermoresponsive Polyisocyanide (PIC) Hydrogels as Bioink and Fugitive Material for Tissue Engineering. *Polymers* **2018**, *10*, 555. [CrossRef]

152. Liu, Y.; Geever, L.M.; Kennedy, J.E.; Higginbotham, C.L.; Cahill, P.A.; McGuinness, G.B. Thermal behavior and mechanical properties of physically crosslinked PVA/Gelatin hydrogels. *J. Mech. Behav. Biomed. Mater.* **2010**, *3*, 203–209. [CrossRef]

153. Moscato, S.; Ronca, F.; Campani, D.; Danti, S. Poly(vinyl alcohol)/gelatin Hydrogels Cultured with HepG2 Cells as a 3D Model of Hepatocellular Carcinoma: A Morphological Study. *J. Funct. Biomater.* **2015**, *6*, 16–32. [CrossRef]

154. Kim, S.S.; Utsunomiya, H.; Koski, J.A.; Wu, B.M.; Cima, M.J.; Sohn, J.; Mukai, K.; Griffith, L.G.; Vacanti, J.P. Survival and function of hepatocytes on a novel three-dimensional synthetic biodegradable polymer scaffold with an intrinsic network of channels. *Ann. Surg.* **1998**, *228*, 8–13. [CrossRef]

155. Kasuya, J.; Sudo, R.; Tamogami, R.; Masuda, G.; Mitaka, T.; Ikeda, M.; Tanishita, K. Reconstruction of 3D stacked hepatocyte tissues using degradable, microporous poly(d,l-lactide-co-glycolide) membranes. *Biomaterials* **2012**, *33*, 2693–2700. [CrossRef]

156. Kobayashi, S.; Nagano, H.; Marubashi, S.; Wada, H.; Eguchi, H.; Takeda, Y.; Tanemura, M.; Doki, Y.; Mori, M. Fibrin sealant with PGA felt for prevention of bile leakage after liver resection. *Hepatogastroenterology* **2012**, *59*, 2564–2568. [CrossRef]

157. Riccalton-Banks, L.; Liew, C.; Bhandari, R.; Fry, J.; Shakesheff, K. Long-Term Culture of Functional Liver Tissue Three-Dimensional Coculture of Primary Hepatocytes and Stellate Cells. *Tissue Eng.* **2003**, *9*, 401–410. [CrossRef]

158. Davis, M.W.; Vacanti, J.P. Toward development of an implantable tissue engineered liver. *Biomaterials* **1996**, *17*, 365–372. [CrossRef]

159. Feng, Z.Q.; Chu, X.H.; Huang, N.P.; Leach, M.K.; Wang, G.; Wang, Y.C.; Ding, Y.T.; Gu, Z.Z. Rat hepatocyte aggregate formation on discrete aligned nanofibers of type-I collagen-coated poly(L-lactic acid). *Biomaterials* **2010**, *31*, 3604–3612. [CrossRef]

160. Lee, J.W.; Choi, Y.J.; Yong, W.J.; Pati, F.; Shim, J.H.; Kang, K.S.; Kang, I.H.; Park, J.; Cho, D.W. Development of a 3D cell printed construct considering angiogenesis for liver tissue engineering. *Biofabrication* **2016**, *8*, 015007. [CrossRef]

161. Grant, R.; Hay, D.C.; Callanan, A. A Drug-Induced Hybrid Electrospun Poly-Capro-Lactone: Cell-Derived Extracellular Matrix Scaffold for Liver Tissue Engineering. *Tissue Eng. Part A* **2017**, *23*, 650–662. [CrossRef]

162. Hashemi, S.M.; Soleimani, M.; Zargarian, S.S.; Haddadi-Asl, V.; Ahmadbeigi, N.; Soudi, S.; Gheisari, Y.; Hajarizadeh, A.; Mohammadi, Y. In vitro differentiation of human cord blood-derived unrestricted somatic stem cells into hepatocyte-like cells on poly(epsilon-caprolactone) nanofiber scaffolds. *Cells Tissues Organs* **2009**, *190*, 135–149. [CrossRef]

163. Semnani, D.; Naghashzargar, E.; Hadjianfar, M.; Dehghan Manshadi, F.; Mohammadi, S.; Karbasi, S.; Effaty, F. Evaluation of PCL/chitosan electrospun nanofibers for liver tissue engineering. *Int. J. Polym. Mater. Polym. Biomater.* **2016**, *66*, 149–157. [CrossRef]

164. Tan, W.J.; Teo, G.P.; Liao, K.; Leong, K.W.; Mao, H.Q.; Chan, V. Adhesion contact dynamics of primary hepatocytes on poly(ethylene terephthalate) surface. *Biomaterials* **2005**, *26*, 891–898. [CrossRef]

165. Janorkar, A.V.; Rajagopalan, P.; Yarmush, M.L.; Megeed, Z. The use of elastin-like polypeptide-polyelectrolyte complexes to control hepatocyte morphology and function in vitro. *Biomaterials* **2008**, *29*, 625–632. [CrossRef]

166. Cho, C.S.; Seo, S.J.; Park, I.K.; Kim, S.H.; Kim, T.H.; Hoshiba, T.; Harada, I.; Akaike, T. Galactose-carrying polymers as extracellular matrices for liver tissue engineering. *Biomaterials* **2006**, *27*, 576–585. [CrossRef]

167. Zhang, Y.; Wang, Q.S.; Yan, K.; Qi, Y.; Wang, G.F.; Cui, Y.L. Preparation, characterization, and evaluation of genipin crosslinked chitosan/gelatin three-dimensional scaffolds for liver tissue engineering applications. *J. Biomed. Mater. Res. A* **2016**, *104*, 1863–1870. [CrossRef]

168. Yan, Y.; Wang, X.; Pan, Y.; Liu, H.; Cheng, J.; Xiong, Z.; Lin, F.; Wu, R.; Zhang, R.; Lu, Q. Fabrication of viable tissue-engineered constructs with 3D cell-assembly technique. *Biomaterials* **2005**, *26*, 5864–5871. [CrossRef]

169. Jiankang, H.; Dichen, L.; Yaxiong, L.; Bo, Y.; Bingheng, L.; Qin, L. Fabrication and characterization of chitosan/gelatin porous scaffolds with predefined internal microstructures. *Polymer* **2007**, *48*, 4578–4588. [CrossRef]

170. Yang, Z.; Xu, L.S.; Yin, F.; Shi, Y.Q.; Han, Y.; Zhang, L.; Jin, H.F.; Nie, Y.Z.; Wang, J.B.; Hao, X.; et al. In vitro and in vivo characterization of silk fibroin/gelatin composite scaffolds for liver tissue engineering. *J. Dig. Dis.* **2012**, *13*, 168–178. [CrossRef]

171. Fan, J.; Shang, Y.; Yuan, Y.; Yang, J. Preparation and characterization of chitosan/galactosylated hyaluronic acid scaffolds for primary hepatocytes culture. *J. Mater. Sci. Mater. Med.* **2010**, *21*, 319–327. [CrossRef]

172. Turner, R.A.; Wauthier, E.; Lozoya, O.; McClelland, R.; Bowsher, J.E.; Barbier, C.; Prestwich, G.; Hsu, E.; Gerber, D.A.; Reid, L.M. Successful transplantation of human hepatic stem cells with restricted localization to liver using hyaluronan grafts. *Hepatology* **2013**, *57*, 775–784. [CrossRef]

173. Dvir-Ginzberg, M.; Gamlieli-Bonshtein, I.; Agbaria, R.; Cohen, S. Liver Tissue Engineering within Alginate Scaffolds Effects of Cell-Seeding Density on Hepatocyte Viability, Morphology, and Function. *Tissue Eng.* **2003**, *9*, 757–766. [CrossRef]

174. Kilbride, P.; Mahbubani, K.T.; Saeb-Parsy, K.; Morris, G.J. Engaging Cold to Upregulate Cell Proliferation in Alginate-Encapsulated Liver Spheroids. *Tissue Eng. Part C Methods* **2017**, *23*, 455–464. [CrossRef]

175. Glicklis, R.; Shapiro, L.; Agbaria, R.; Merchuk, J.C.; Cohen, S. Hepatocyte behavior within three-dimensional porous alginate scaffolds. *Biotechnol. Bioeng.* **2000**, *67*, 344–353. [CrossRef]

176. Yang, J.; Chung, T.W.; Nagaoka, M.; Goto, M.; Cho, C.-S.; Akaike, T. Hepatocyte-specific porous polymer-scaffolds of alginate-galactosylated chitosan sponge for liver-tissue engineering. *Biotechnol. Lett.* **2001**, *23*, 1385–1389. [CrossRef]

177. Wang, B.; Hu, Q.; Wan, T.; Yang, F.; Cui, L.; Hu, S.; Gong, B.; Li, M.; Zheng, Q.C. Porous Lactose-Modified Chitosan Scaffold for Liver Tissue Engineering: Influence of Galactose Moieties on Cell Attachment and Mechanical Stability. *Int. J. Polym. Sci.* **2016**, *2016*, 1–8. [CrossRef]

178. She, Z.; Jin, C.; Huang, Z.; Zhang, B.; Feng, Q.; Xu, Y. Silk fibroin/chitosan scaffold: Preparation, characterization, and culture with HepG2 cell. *J. Mater. Sci. Mater. Med.* **2008**, *19*, 3545–3553. [CrossRef]

179. She, Z.; Zhang, B.; Jin, C.; Feng, Q.; Xu, Y. Preparation and in vitro degradation of porous three-dimensional silk fibroin/chitosan scaffold. *Polym. Degrad. Stab.* **2008**, *93*, 1316–1322. [CrossRef]

180. Lewis, P.L.; Shah, R.N. 3D Printing for Liver Tissue Engineering: Current Approaches and Future Challenges. *Curr. Transplant. Rep.* **2016**, *3*, 100–108. [CrossRef]

181. Miller, J.S.; Stevens, K.R.; Yang, M.T.; Baker, B.M.; Nguyen, D.H.; Cohen, D.M.; Toro, E.; Chen, A.A.; Galie, P.A.; Yu, X.; et al. Rapid casting of patterned vascular networks for perfusable engineered three-dimensional tissues. *Nat. Mater.* **2012**, *11*, 768–774. [CrossRef]

182. Gou, M.; Qu, X.; Zhu, W.; Xiang, M.; Yang, J.; Zhang, K.; Wei, Y.; Chen, S. Bio-inspired detoxification using 3D-printed hydrogel nanocomposites. *Nat. Commun.* **2014**, *5*, 3774. [CrossRef]

183. Underhill, G.H.; Chen, A.A.; Albrecht, D.R.; Bhatia, S.N. Assessment of hepatocellular function within PEG hydrogels. *Biomaterials* **2007**, *28*, 256–270. [CrossRef]

184. Hammond, J.S.; Gilbert, T.W.; Howard, D.; Zaitoun, A.; Michalopoulos, G.; Shakesheff, K.M.; Beckingham, I.J.; Badylak, S.F. Scaffolds containing growth factors and extracellular matrix induce hepatocyte proliferation and cell migration in normal and regenerating rat liver. *J. Hepatol.* **2011**, *54*, 279–287. [CrossRef]

185. Stevens, K.R.; Miller, J.S.; Blakely, B.L.; Chen, C.S.; Bhatia, S.N. Degradable hydrogels derived from PEG-diacrylamide for hepatic tissue engineering. *J. Biomed. Mater. Res. A* **2015**, *103*, 3331–3338. [CrossRef]

186. Zimoch, J.; Padial, J.S.; Klar, A.S.; Vallmajo-Martin, Q.; Meuli, M.; Biedermann, T.; Wilson, C.J.; Rowan, A.; Reichmann, E. Polyisocyanopeptide hydrogels: A novel thermo-responsive hydrogel supporting pre-vascularization and the development of organotypic structures. *Acta Biomater.* **2018**, *70*, 129–139. [CrossRef]

187. Hasirci, V.; Berthiaume, F.; Bondre, S.P.; Gresser, J.D.; Trantolo, D.J.; Toner, M.; Wise, D.L. Expression of Liver-Specific Functions by Rat Hepatocytes Seeded in Treated Poly(Lactic-co-Glycolic) Acid Biodegradable Foams. *Tissue Eng.* **2001**, *7*, 385–394. [CrossRef]

188. Wang, X.; Rijff, B.L.; Khang, G. A building-block approach to 3D printing a multichannel, organ-regenerative scaffold. *J. Tissue Eng. Regen. Med.* **2017**, *11*, 1403–1411. [CrossRef]

189. Bishi, D.K.; Mathapati, S.; Venugopal, J.R.; Guhathakurta, S.; Cherian, K.M.; Verma, R.S.; Ramakrishna, S. A Patient-Inspired Ex Vivo Liver Tissue Engineering Approach with Autologous Mesenchymal Stem Cells and Hepatogenic Serum. *Adv. Healthc. Mater.* **2016**, *5*, 1058–1070. [CrossRef]

190. Huang, H.; Oizumi, S.; Kojima, N.; Niino, T.; Sakai, Y. Avidin-biotin binding-based cell seeding and perfusion culture of liver-derived cells in a porous scaffold with a three-dimensional interconnected flow-channel network. *Biomaterials* **2007**, *28*, 3815–3823. [CrossRef]

191. Rad, A.T.; Ali, N.; Kotturi, H.S.; Yazdimamaghani, M.; Smay, J.; Vashaee, D.; Tayebi, L. Conducting scaffolds for liver tissue engineering. *J. Biomed. Mater. Res. A* **2014**, *102*, 4169–4181. [CrossRef]

192. Blouin, A.; Bolender, R.P.; Weibel, E.R. Distribution of organelles and membranes between hepatocytes and nonhepatocytes in the rat liver parenchyma. A stereological study. *J. Cell Biol.* **1977**, *72*, 441–455. [CrossRef]

193. Page, D.T.; Garvey, J.S. Isolation and characterization of hepatocytes and Kupffer cells. *J. Immunol. Methods* **1979**, *27*, 159–173. [CrossRef]

194. Lee, H.; Han, W.; Kim, H.; Ha, D.H.; Jang, J.; Kim, B.S.; Cho, D.W. Development of Liver Decellularized Extracellular Matrix Bioink for Three-Dimensional Cell Printing-Based Liver Tissue Engineering. *Biomacromolecules* **2017**, *18*, 1229–1237. [CrossRef]

195. Mazza, G.; Al-Akkad, W.; Rombouts, K.; Pinzani, M. Liver tissue engineering: From implantable tissue to whole organ engineering. *Hepatol. Commun.* **2018**, *2*, 131–141. [CrossRef]

196. Ambekar, R.S.; Kandasubramanian, B. Progress in the Advancement of Porous Biopolymer Scaffold: Tissue Engineering Application. *Ind. Eng. Chem. Res.* **2019**, *58*, 6163–6194. [CrossRef]

197. Lewis, P.L.; Green, R.M.; Shah, R.N. 3D-printed gelatin scaffolds of differing pore geometry modulate hepatocyte function and gene expression. *Acta Biomater.* **2018**, *69*, 63–70. [CrossRef]

bioengineering

MDPI

Review

Bioprinting for Liver Transplantation

Christina Kryou, Valentina Leva, Marianneza Chatzipetrou and Ioanna Zergioti *

Department of Physics, National Technical University of Athens, 15780 Zografou, Greece;
chkryou@central.ntua.gr (C.K.); vleva@mail.ntua.gr (V.L.); mchatzip@mail.ntua.gr (M.C.)
* Correspondence: zergioti@central.ntua.gr; Tel.: +30-210-772-3345

Received: 16 July 2019; Accepted: 25 September 2019; Published: 10 October 2019

Abstract: Bioprinting techniques can be used for the in vitro fabrication of functional complex bio-structures. Thus, extensive research is being carried on the use of various techniques for the development of 3D cellular structures. This article focuses on direct writing techniques commonly used for the fabrication of cell structures. Three different types of bioprinting techniques are depicted: Laser-based bioprinting, ink-jet bioprinting and extrusion bioprinting. Further on, a special reference is made to the use of the bioprinting techniques for the fabrication of 2D and 3D liver model structures and liver on chip platforms. The field of liver tissue engineering has been rapidly developed, and a wide range of materials can be used for building novel functional liver structures. The focus on liver is due to its importance as one of the most critical organs on which to test new pharmaceuticals, as it is involved in many metabolic and detoxification processes, and the toxicity of the liver is often the cause of drug rejection.

Keywords: additive manufacturing; direct printing; 3D structuring; tissue engineering

---

1. Introduction

Over the past few decades, printing technology has advanced from two-dimensional (2D) printing to an additive process in which successive layers of material are arranged to create 3D objects [1,2]. The ability of printing techniques to produce 3D structures with complex geometries and structures enables rapid prototyping and manufacturing in the industry, as well as the production of personalized medicine.

The 3D printing field was first introduced in 1986 by Charles W. Hull as "stereolithography" [3]. In this technique, thin layers of a material were printed in layers to form solid 3D structures using photochemical processes. Since the 1990s, stereolithographic models have been used for creating sacrificial resin molds for the formation of 3D scaffolds of biological materials. Those materials are used for transplantation with or without seeded cells [4]. The next generation was "3D bioprinting", which was used as a tool for tissue engineering and organ fabrication.

3D bioprinting employs the controlled, precise delivery and placement of living cells, biomaterials and biochemicals to fabricate functional 3D constructs in a layer by layer manner. 3D bio-printing has emerged as one of the most influential applications of 3D printing, aiming to address the increased demand for living constructs with long term mechanical and biological stability, suitable for transplantation and improved drug discovery models [5,6]. 3D bio-printing permits rapid manufacturing with high-precision and control over size, as well as adjustments to the shape, porosity, and mechanical strength of the scaffolds in one step; it has thus attracted much attention in the tissue engineering field. One of the main drawbacks of the 3D bioprinting technologies is the vascularization of the created tissue structure, which still remains a critical challenge. The development of vascular networks within densely populated and metabolically functional tissues facilitate the transport of nutrients and oxygen, and it provides a way to remove wastes, for which the long term preservation of cellular viability can be obtained. Moreover, it has been considered as a promising method to replace defective or damaged tissues or organs in which scaffolds have functioned as carriers for cell interaction and provided physical support to the freshly developed tissue [7].

Impressive progress has been accomplished in fabricating complex tissue constructs in the past few years. The main approaches for controlled 3D vascularization within these engineered tissues mainly involve microfluidic-based technologies [8,9]. The microfluidic-based technologies can provide a versatile platform for engineered tissues because they can create complex and functional micro-scale environments in order to mimic 3D in vivo environments (e.g., a chemical gradient). Moreover, microfluidic technologies have emerged as useful tools for complex cell environments like tissues due to the integration of multiple steps and fluid control, such as controllable cell culture, cell capture, mixing, genetic assays, protein and continuous nutrition, and oxygen supply [10–13]; however, they are also limited by fabrication complexity. A functional circulatory system is a key factor for the creation of tissue constructs which are limited to a distance of just a few hundred microns but are not limited to diffusion for nutrition [14]. In addition, innovative strategies such as the guided infiltration of host micro vessels into the implanted construct, the integration of autologous vascular grafts, and the direct bioprinting of vascular structures have also been attempted by the research community [15].

This review aims to highlight the techniques used for the patterning of cells towards the creation of a structures with increased complexity such as tissues and organs. Special attention is given to the techniques used for the fabrication of tissue structures such as the creation of 3D scaffolds and/or direct printing techniques, as well as the combination of both approaches.

2. 3D Bioprinting Techniques

The liver is an extremely important organ for functions related to metabolism and metabolic regulation. Unfortunately, liver failure or acute chronic liver failure remains one of the most major causes of mortality in the world. As a result of the increase in liver diseases, the need for donor organs is increasing [16]. Despite the great importance of the organ in a human's life, liver transplantation is usually performed only on patients with major and/or end-stage liver diseases due to the short life span of donor organs or rejection risk. Consequently, alternative methods, including tissue engineering, are needed and are actively being pursued. The field of liver tissue engineering includes several techniques aimed at providing therapeutic development for liver diseases and plays an important role in the mechanistic understanding of liver biology interactions in healthy and diseased states in a high throughput platform. Artificial liver transplantation is a recent challenge in medicine, as it has been deemed the best therapeutic method for severe liver diseases. Conventional liver tissue models have recently been used to fabricate in vitro 3D liver tissue models [17]. These methods can be classified into four main categories: (i) Monolayer cell cultures, including aggregating and assembly techniques; (ii) hollow fiber; (iii) suspension chambers; and (iv) perfusion beds [18]. Nevertheless, these approaches often fail to imitate the complexity of native liver tissue and are incapable of depositing multiple cell types in desired patterning [19].

Three-dimensional (3D) printing, which belongs to the family of additive manufacturing techniques [20], can resolve issues inherent to traditional 2D and 3D models, such as the low efficacy of engraftment and poor cellular functions, because 3D printing provides the ability to manipulate cell–cell interactions as opposed to conventional models.

3D printing was first developed in the 1980s, and there have been enormous advancements in tissue and organ regeneration [21]. In 1993, the first 3D printer was designed by Sachs et al. to print nonviable materials, such as plastics and metals [22]. Since then, a number of 3D printers [23] have been successfully designed and used for tissue biofabrication and regenerative medicine [5,24] Typically, 3D bioprinting starts with a computer-aided process for depositing biological materials such as living cells, matrices, biomaterials, and molecules in a layer by layer manner with a prescribed configuration in order to produce scalable bioengineered structures [25]. In this way, 3D biomimetic tissue models with heterogeneous cell placements and vasculature have been proposed as means to recapitulate liver tissue complexity and architecture [26]. The fabrication of perfectly functional liver networks remains a challenge for most tissue engineers. Hence, there are considerable types of 3D printing methods that are expected to overcome current limitations. 3D bioprinting offers the ability to develop highly complex 3D patterns with living cells that mimic organ level functions, and it has therefore been applied in organs-on-chips and organs engineering. The main bioprinting techniques are extrusion [27], inkjet [28] and laser-induced forward transfer (LIFT) [29,30], each one possessing several advantages and disadvantages.

2.1. Laser Bioprinting

Laser-induced forward transfer (LIFT) is a technique presented more than 30 years ago by Bohandy et al. [31]. Briefly, a pulsed laser beam is applied on a donor slide (or ribbon) covered with a laser-energy-absorbing layer (e.g., gold or titanium) containing the desired material (e.g., cells, hydrogels and growth factors), followed by the evaporation of the material; this results in a high-pressure bubble jetting toward the receiving substrate that is placed underneath the donor slide, as shown in Figure 1.

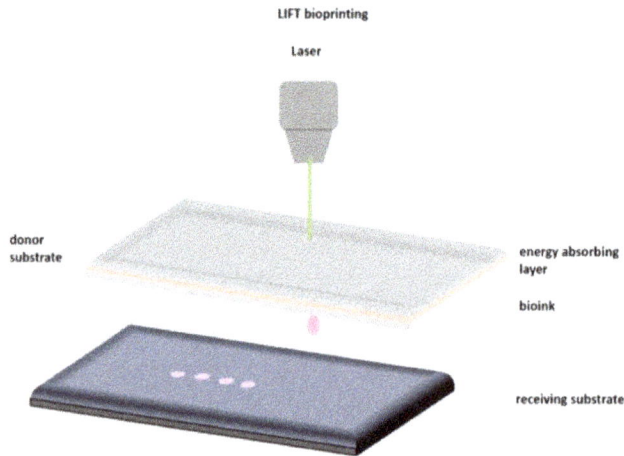

Figure 1. Schematic representation of laser-induced forward transfer (LIFT) setup.

For the direct printing of cells, the use of LIFT is proposed because it enables the printing of bio-inks within a wide range of viscosities (1–300 mPa s) [32] and at high speeds while cell viability is preserved (>90%).

LIFT is a nozzle-free technique and therefore does not have the problems of nozzle clogging with cells or biological materials, which are some major drawbacks of other bioprinting technologies. Moreover, this technique offers printing cell concentrations up to 1×108 cells/mL with a very high resolution [33].

The use of LIFT for the printing of functional biomaterials can be traced back to 2003 [34], while the development of 2D cell structures was first proposed in 2008 [35]. Regarding the use of lasers for the printing of 3D structures, the first report was published in 2011 by M. Gruene et al. [36], while in 2012, Koch et al. [37] published the printing of multiple cell lines in order to create epidermal tissue. These multiple cell lines were previously proven to be resistant to damage during the laser-assisted printing process [38]. The proliferation of cells over a period of 10 days was studied, and the ability of 3D printed cells to form real tissue was demonstrated. It is critical to know how the laser process affects cell viability as well as phenotypes. Catros et al. [39] studied the effects of laser pulse energy, extracellular matrix (ECM) thickness and viscosity of the bioink on cell viability. Cell viability 24 h post-printing was measured to compare different printing settings. It was concluded that while higher laser energy leads to more cell fatality, increasing film thickness as well as bioink viscosity results in increased cell viability. Moreover, another laser group investigated the effects of bioink viscosity, laser energy and printing speed on printing resolution [32]. It was shown that a microscale resolution and 5 kHz printing speed were within reach. This work is another proof for the applicability of printing cells and biomaterials via LIFT printing to engineer miniaturized tissue layouts with de novo high cell density and microscale organization. An interesting study was demonstrated by Keriquel et al. [40], whereby in vivo laser bioprinting was used to deposit nano-hydroxyapatite in a mouse calvaria 3D defect model as a proof of concept. In the future, study materials that can directly integrate into a patient's tissue could be used. Finally, incorporating the patients' own cells may facilitate the applicability of these types of constructs to contribute to both the structural and functional components of the tissue.

2.2. Inkjet Bioprinting

Inkjet-based bioprinting is a noncontact technique in which droplets of cells or biomaterials are patterned into desired substrates.

The drop-on demand inkjet bioprinters are the most common ones, and they consist of thermal, piezoelectric, and electrostatic inkjet nozzles [41]. A schematic diagram of inkjet printing is shown in Figure 2.

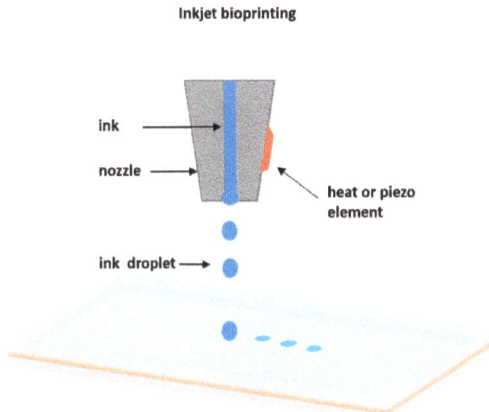

Figure 2. Schematic representation of inkjet printing.

With respect to the construction of cellular structures, inkjet bioprinters are normally used for the printing of matrices for the cell growth, such as small scaffolds. Different inkjet printheads with multiple nozzles have been developed to increase printing speed and fabricate larger cellular constructs [42].

However, inkjet bioprinters also have limitations on material viscosity (ideally below 10 centipoise) due to the excessive force required to eject drops using solutions at higher viscosities [43]. Another major disadvantage of this technique is the difficulty in achieving biologically relevant cell densities. Often, low cell concentrations are used to facilitate droplet formation (less than 10 million cells/mL) [44]. To provide a higher concentration of cells, the inhibition of some hydrogels can be generated by adding crosslinking agents. However, the requirement for crosslinking agents often slows the bioprinting process and involves the chemical modification of naturally occurring ECM materials, which changes both their chemical and material properties [45]. Despite these disadvantages, inkjet bioprinting has notable benefits, including low cost, high speed and biocompatibility with a broad range of biological materials [46]. Significant studies of inkjet printing have included the regeneration of functional tissues, such as skin and cartilage, in situ [47,48]. With the advantages of high throughput digital control and high resolution, this technique enables the direct placement of cells, biological factors and biomaterial scaffolds directly into skin or cartilage lesions. Inkjet-based bioprinting facilitates the successful deposition of either primary cells or stem cell types with uniform density, and it maintains high cell viability and function after printing. These studies have shown the ability of inkjet bioprinting to regenerate functional constructs.

2.3. Extrusion Bioprinting

The extrusion-based bioprinting technique is characterized by a temperature-controlled biomaterial dispensing system driven by a pneumatic pressure or a mechanical piston, as demonstrated in Figure 3. Schematic representation of extrusion bioprinting. The printing system generates continuous biomaterial filaments, instead of droplets, that are deposited in two dimensions; filaments are placed along the *x*- and *z*-axes and then move higher in the *y*-axis. The final product is a 3D structure. This technique provides the ability to deposit very high cell densities as well biological material such as hydrogels and biocompatible copolymers. Several groups have used sole cells or multicellular cell spheroids and allowed for their self-assembly into the desired 3D structures using extrusion bioprinters [49–51]. Pioneer work using this approach is currently being performed at the Wyss Institute under Prof J. Lewis [52]. Each print head is equipped with an on-board temperature controller to adjust the temperature depending on the material that is being printed, enabling sequential layer-by-layer printing and avoiding contamination between different materials.

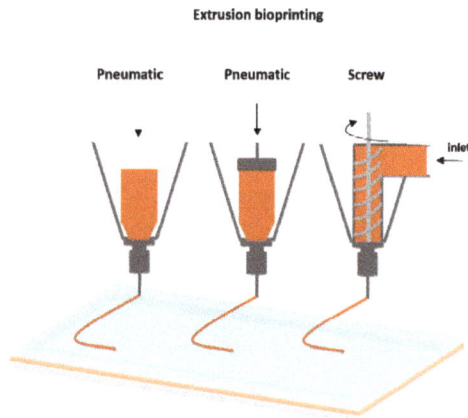

Figure 3. Schematic representation of extrusion bioprinting.

However, a major disadvantage of extrusion bioprinting is that cell viability is lower than that with inkjet-based bioprinting (40–86%). The decreased cell survival rate possibly results from the shear stresses inflicted on cells in viscous fluids [53].

Extrusion-based bioprinting approaches have been also used for the generation of multiple tissue types, including aortic valves [54] and in vitro pharmokinetic models [55].

A review of the outstanding research works using the above printing techniques for liver and liver tissue engineering is presented.

A brief review of the above mentioned bioprinting techniques is presented in Table 1. A brief review of common bioprinting techniques.

Table 1. A brief review of common bioprinting techniques.

	Laser Assisted Bioprinting	Inkjet	Extrusion
Advantages	High resolution, deposition of biomaterials in solid or liquid phase, and nozzle free and non-contact printing.	Ability to print low viscosity biomaterials, fast fabrication speed, low cost, high resolution, multi-material printing, Simple operation.	Simple, capable of printing various biomaterials, ability to print high cell densities, multi-material printing, and ability to control ejection speed.
Drawbacks	High cost, thermal damage due to nanosecond/femtosecond laser irritation, metallic residuals possible damage of tissue from use of laser lights, slow printing speed, and difficulty in handling heterogenous cells.	Inherent inability to provide a continuous flow, poor functionality for vertical structures, low cell densities, clogging of nozzle, imposing thermal or acoustic stress to cells, and limited variety of bioink.	Only applicable for viscous liquids, gelation and solidification, and limited material selection (shear thinning ability required).
Speed	Medium	Fast	Slow
Cell viability	<85%	~80%	>90%
Resolution	10 μm	50 μm	100 μm
Cell density	Medium	Low	High
Viscosity	1–300 mPa s	<10 mPa s	30–6×10^7 mPa s
Scalability	Low	Low	Low–Medium
Structural integrity	Low	Low	High
Cost	High	Low	Low–Medium

3. Tissue and Liver Bioprinting

As previously mentioned, the liver is considered one of the most significant organs in the human body due to its special characteristics. It plays a major role in metabolism with numerous functions, including the regulation of glycogen storage, the decomposition of red blood cells, plasma protein synthesis, hormone production, and the detoxification of chemicals [56,57]. In anatomy, the liver is divided into four lobes. The right lobe, which is much bigger than the left lobe, involves two minor lobes—the quadrate and caudate lobes. Blood is supplied to the liver through two different vessels. The hepatic artery supplies arterial blood from the heart to the liver, and the hepatic portal vein carries blood consisting of nutrients and toxins from the intestines to the liver [57].

The liver has an extensive regeneration capacity due to the high proliferation ability of hepatocytes, even if it is subjected to vast damages. The tissue engineering of the liver is not new, and there are several groups that have worked on the engineering of liver tissues and bioartificial livers as early as 1996 [58,59]. Therefore, various tissue bioprinting techniques have been used to fabricate biomimetic liver tissues—even a whole liver. A schematic representation of the key approaches used for liver tissue engineering is demonstrated in Figure 4 [59].

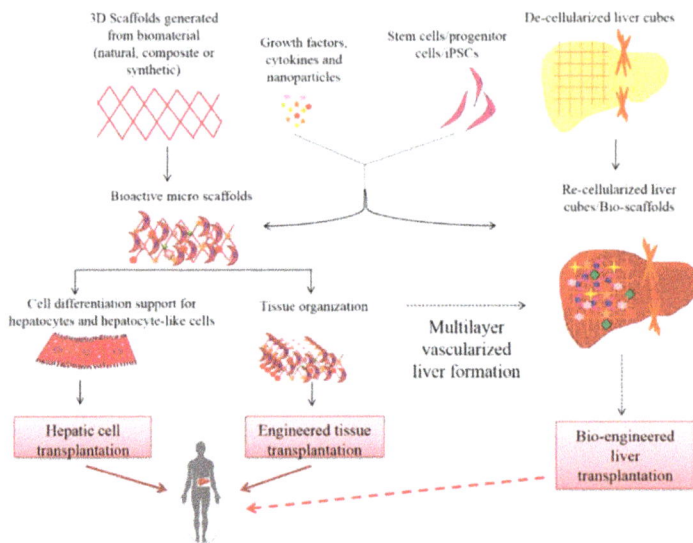

Figure 4. Schematic diagram of liver tissue engineering. Solid lines show already ongoing approaches, whereas dotted lines indicate proposed mechanisms [59].

3.1. Micropatterned 2D and 3D Liver Models

Over the past few decades, liver tissue engineering has made significant progress towards the establishment of in vitro liver models for both fundamental pathophysiological studies and drug screening. The sources of cells used for these in vitro liver models include primary hepatocytes, hepatic cell lines isolated from tumors or liver slices, and stem cell-derived hepatic cells [60,61]. Griffith et al. [62] fabricated a vascularized liver on a small scale using the inkjet printing technique. They were pioneers in investigating the role of scaffold architecture from biodegradable polyesters using a manufacturing technique amenable to scaling-up, commercial production, and culture conditions for achieving hepatic function in long-term perfusion cultures.

Monolayer culture, organoid culture and co-culture platforms have been established using culture plates [63], commercially available wells [64], dielectrophoresis micropatterning [65] and physical mask-based additive photopatterning methods [60]. However, the liver specific functions of hepatocytes cultured in such platforms are functional only for weeks of in vitro culture [63,66]. Therefore, liver constructs that better mimic the native environment and help maintain in vitro liver functions is in great demand.

3D bioprinting technology, with its potential to pattern cells and biomaterials in a precise manner, provides a great tool to achieve novel and biomimetic in vitro liver models with increasing structural complexity.

3.2. 3D Bioprinting for Liver Models

3D printing is a scientific field with innovative techniques that offer remarkable benefits in terms of the vascular network formation of liver tissues and organs due to their feasibility, variety of available printing methods, and precise controllability. With the appearance of bioprinting, the constructions of functional tissue livers or mini liver organs have become an impending reality. Currently, many researchers are contributing to the improvement of 3D printed vascular networks on a best effort basis for their introduction into the medical field.

Many researchers that have worked on tissue engineering have successfully achieved to fabricate biomimetic 3D printed vascularized liver constructs with their own unique properties such as rapid restoration ability even after considerable damage [67]. In an earlier work by Cheng et al. [68], 30 layers of a hepatocyte/gelatin mixture were laminated into a high spatial structure using a 3D rapid prototyping technology. The 3D hepatocyte/gelatin pattern remained viable and performed biological functions in the construct for more than two months. In an effort to develop personalized tissues and organs for precision medicine, Organovo, harnessing the advantages of 3D bioprinting, used a syringe-based extrusion printer to develop 3D printed human liver tissues that can remain fully functional and stable for up to 28 days. The researcher demonstrated a multicellular liver structure involving hepatocytes, hepatic stellates, and endothelial cells (ECs). 3D liver tissues possessed critical liver functions, including albumin production, cholesterol biosynthesis, fibrinogen and transferrin production, and inducible cytochrome (CYP) 1A2 and CYP 3A4 activities. These in vitro models of 3D vascularized livers could potentially be implanted into patients to replace their damaged livers [69]. In 2013, the first human liver was synthetically reproduced and validated against the actual native liver at the time of surgery by Zein et al. [70]. Specifically, successful 3D synthetic livers were printed along with their complex network of vascular and biliary structures which replicated the native livers for six patients, three living donors, and three respective recipients. Prior to the transplantation, the dimensions of the donor and recipient livers were recorded in detail, including the diameters of veins to fabricate a vascularized liver using the inkjet printing technique and based on each patient's individual computed tomography (CT) scan and magnetic resonance imaging (MRI). To implement external vascularization, the authors utilized a permanent adhesive to attach to the liver lobe (Figure 5). These results demonstrate the potential efficacy of a 3D printed synthetic liver with a vascular network in the human body as a valuable tool for drug delivery, a substitute for treating partially or irreversibly damaged liver tissue, and a tool for potentially minimizing intraoperative complications. That was the first human liver to have been synthetically reproduced and validated against the actual native liver at the time of surgery.

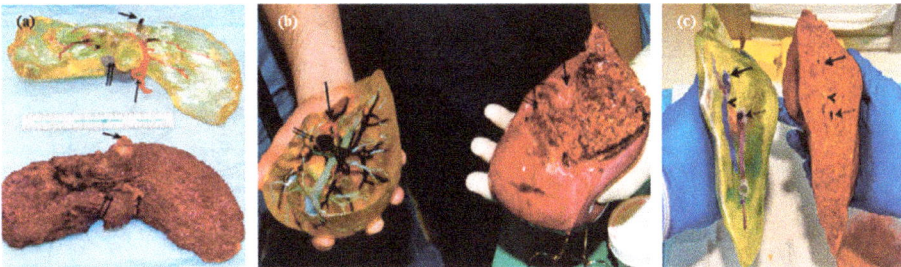

Figure 5. (a) Side view of a 3D printed liver and extracted liver of a patient, where long, short, and double arrows indicate hepatic artery, hepatic vein, and portal vein, respectively. (b) Right lobes of 3D printed and extracted livers with indications of the hepatic artery (single arrows) and portal vein (double arrows). (c) Cross-sectional views of 3D printed and extracted livers with indications of hepatic vein (single arrows) and portal vein (dotted arrows) [70].

Nguyen et al. [71], established a novel bioprinted human mini liver tissue from the co-culture of primary human hepatocytes, hepatic stellate cells (HSC) and human umbilical vein endothelial cells (HUVEC) cells to test clinical drug-induced toxicity in vitro using an inkjet 3D bioprinter. A histological analysis showed the presence of distinct intercellular hepatocyte junctions, cluster of differentiation 31 (CD31+) endothelial networks, and desmin-positive, smooth muscle actin-negative quiescent stellates, mimicking the in vivo human drug response at the tissue level (Figure 6). A major challenge in liver tissue engineering is the proliferation, long-term culture and maintenance of hepatocyte function ex vivo of primary hepatocytes [38].

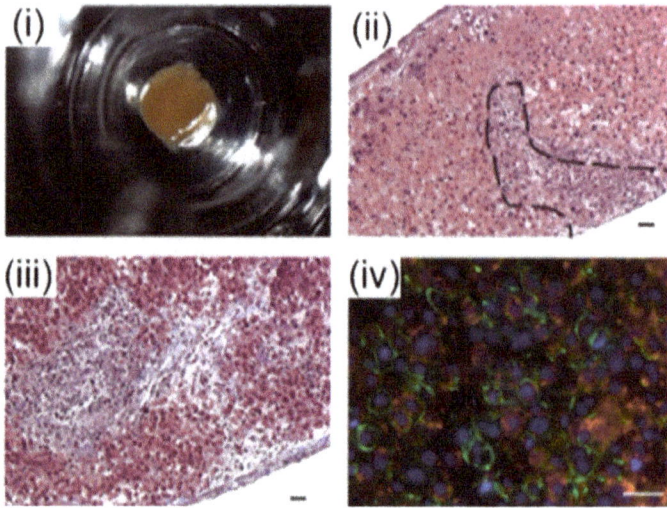

Figure 6. Organovo's mini liver tissue: (**i**) A macroscopic image of liver tissue housed in a 24-well transwell, (**ii**) Hematoxylin and eosin (HE) staining of a tissue cross-section, (**iii**) extracellular matrix (ECM) deposition assessed by Masson's trichrome staining, and (**iv**) Iimmunohistochemistry (IHC) staining of the parenchymal compartment for E-cadherin (green) and albumin (red) [71].

A recent study by our team [72] utilized the LIFT technique to laser print hepatocyte cancer cell line Huh7 on porous collagen-Glycosaminoglycan (GAG) scaffolds, which are biomaterials with established applications in re-generative medicine implants. The results showed the benefits of the laser bioprinting technique for the precise placement and immobilization of hepatocyte cells into porous collagen scaffolds for novel custom-made implants for regenerative medicine applications.

Arai et al. [73] used an inkjet 3D bioprinter to fabricate a 3D culture system using an artificial scaffold for studying the liver-specific functions of hepatocytes. The printed construct expressed liver-specific proteins and receptors such as albumin, MPR2, and asialoglycoprotein receptor (ASGPR), thus proving the functionality of the printed liver tissue. The work by Matsusaki et al. [74] demonstrated that high cell activities and high cell–cell interactions of the fabricated 3D human liver chip from HepG2/HUVECs laden fibronectin and gelatin using inkjet printing technology were analogous to the native liver structure due to the hierarchical sandwich structures.

In another study by Y Kim et al. [75], mouse primary hepatocytes (isolated from the livers of six-to-eight weeks old mice) were printed into a 3D liver tissue construct using an extrusion-based bioprinting system. Cells were viable for 14 days, with liver-specific gene expressions, namely albumin, hepatocyte nuclear factor 4 alpha (HNF-4α), forkhead box protein A3 (Foxa3), and asialoglycoprotein receptor 1 (ASGR1), increasing gradually up to day 14. In another study, Lee et al. [76] developed 3D structures from polycaprolactone (PCL) with improved mechanical properties for liver tissue regeneration by using a multi-head tissue building printing system. A co-cultured 3D microenvironment of primary rat hepatocytes (HCs), human umbilical vein endothelial cells (HUVECs), and human lung fibroblasts (HLFs) were successfully established and maintained to study liver cells proliferation. The results of this work suggested that the employed co-cultured microenvironment promoted heterotypic cellular interaction within a 3D construct. Similarly, Skardal et al. [77] utilized a 3D bioprinting platform to fabricate liver tissue, which has high potential for influencing how future drug and toxicology screening and personalized medicine approaches are performed. Measurable levels of both albumin and urea as well as common soluble biomarkers for liver were tested, and these remained relatively consistent throughout the culture period. Moreover, this group developed a 3D liver tissue model containing primary human hepatocytes and liver stellate cells supported by bioinks, and they tested the functional indicators. Specifically, these constructs were maintained in culture for six days, and liver functionality was examined by exposing the constructs to a hepatic toxicant, acetaminophen (APAP, 100 μM), and measuring the levels of albumin, urea, α-GST (alpha Glutathione S-Transferase), and lactic acid dehydrogenase (LDH) in the media over time. An analysis of both urea and albumin levels showed a significant decrease until day 15 for the acetaminophen-treated conditions. In addition, the levels of α-GST, a detoxification protein, increased at day nine, and the levels of lactic acid dehydrogenase (LDH), a marker of liver damage, also peaked due to printing-related stress but decrease to nominal levels by day six. Finally, histological staining presented a greater cellularity in untreated constructs, while drug-treated conditions showed a loss of cellularity. In the future, these models could be used for drug screening, disease modeling, and precision medicine applications [78]. An interesting decellularized extracellular matrix (dECM) bioink derived from a native liver was demonstrated by Lee et al. [79]. The proposed bioink, in combination with the 3D bioprinting technology, could be a suitable biomechanical and biochemical microenvironment for liver tissue function. Specifically, the cell-printed mixtures consisted of dECM bioink seeded with human bone marrow-derived mesenchymal stem cells (BMSCs) and liver cancer cells (human hepatocellular carcinoma), as well as PCL polymer for 3D structural support, with control constructs prepared with a collagen bioink. The resulting cell-laden printed bioink was evaluated and compared with those in commercial collagen bioink. An analysis of liver-specific functions of these constructs by assessing albumin and urea levels presented that the dECM bioink enhanced liver cell functions. Moreover, the level expression of key transcription factor HNF4A (Hepatocyte nuclear factor 4 alpha) was particularly upregulated in the liver dECM group to more than twice the level seen for the collagen, and the level expression of transcriptional markers HNF1A and HNF3B (Hepatocyte nuclear factor 3-beta) was significantly higher in the liver dECM group.

A recent study by Kurreck et al. [80] utilized the extrusion bioprinting technique to print a 3D tissue model composed of bioinks and human bipotent hepatic progenitor cells (HepaRG) with established applications in virus biology. A short summary of recent outstanding bioprinting studies is presented in Table 2.

Table 2. A short summary of outstanding recent liver bioprinting studies.

Printing Method	Cell Type/Bioink	Achievements	Reference
Extrusion bioprinting	Hepatocytes Gelatin	The laminated hepatocytes remained viable and performed biological functions for more than 2 months	[68]
Extrusion-based bioprinting	Primary human hepatocytes, hepatic stellates, HUVEC cells, and non-parenchymal cells/NovoGelR 2.0 hydrogel (concentration not mentioned)	Viable up to 28 days (% not mentioned) Inkjet bioprinting Galactosylated alginate (12 mg/mL) Primary mouse hepatocytes (isolated from the liver tissue of male 6–8-weeks-old ICR 12 mice) Data not available >85% after 2 Days test of hepatotoxicity of trovafloxacin and Levofloxacin	[71]
Inkjet bioprinting	Primary mouse hepatocytes (isolated from the liver tissue of male 6-to-8-week-old ICR 12 mice)/Galactosylated alginate (12 mg/mL)	>85% after 2 days	[73]

Table 2. *Cont.*

Printing Method	Cell Type/Bioink	Achievements	Reference
Inkjet bioprinting	HUVEC	Multilayered organ tissue model test of hepatotoxicity of troglitazone (Rezulin)	[74]
Extrusion-based bioprinting	Primary mouse hepatocytes (isolated from the livers of 6–8 weeks old mice)/Alginate (3% *w/v*)	Viable up to 14 days (% not mentioned)	[75]
Extrusion bioprinting	HepG2, BMMSCs/decellularized extracellular matrix (dECM)	Liver tissue model	[79]
Microvalve bioprinting	hiPSCs (human-induced pluripotent stem cell lines, RCi-22 and RCi-50); hESCs human embryonic stem cell lines, RC-6 and RC-10)/Alginate (1.5% *w/v*)	>55% after 1 day	[81]
Extrusion-based bioprinting	Primary hepatocytes	Viable up to 60 days (% not mentioned)	[82]

An alternative approach to liver tissue fabrication is the use of stem cells. Concerning the hepatic differentiation of induced pluripotent stem cells (iPSCs) to liver-specific cell lines. The first successful work on bioprinting a mini-liver from both human-induced pluripotent stem cells (hiPSCs) and human embryonic stem cells (hESCs), which have matured to be hepatocyte-like cells, was reported by Faulkner-Jones et al. using a valve-based bioprinting system which did not adversely affect cell viability (~84%) [83]. The group built a 3D alginate matrix, and the analysis was carried out after 21 days of differentiation protocol, revealing peak albumin secretion that meant the construct was hepatic in nature [81], as shown in Figure 7 [84]. Recently, Choi et al. [85] used a nozzle 3D bioprinter to fabricate a liver-mimicking architecture using primary hepatocytes, and they demonstrated the benefits of co-cultured primary hepatocytes and mesenchymal stem cells (MSCs). This research indicated that the expression of hepatic genes and proteins was higher for up to seven days in the 3D hepatic architecture, and that the primary hepatocyte cell morphology was stable.

Most 3D-bioprinted tissues demonstrate liver-specific functions in addition to injury response. Several companies and research groups have created living constructs that mimic native liver structures and functions [86–89].

There is an acute demand for livers, and the fabrication of liver tissue or liver will definitely alleviate this problem. Liver tissue and organoids can also be used in other assays such as drug testing and liver disease studies. As with mature hepatocytes, hepatocyte-like cells obtained from stem cells tend to quickly functionally deteriorate under in vitro conditions. The liver structure is complex with a modular microenvironment; thus, it is difficult to model native liver tissue [87]. Recently, Kizawa et al. [82] printed a liver tissue by the spheroid assembly of primary hepatocytes (1×10^4 cells/mL) that maintained functionality up to 60 days by using a scaffold-free 3D bioprinting technology from Cyfuse Biomedical (NA1002, Cyfuse Biomedical), as demonstrated in Figure 8. The human 3D bioprinted liver construct also maintained the expression of many drug transporter proteins and metabolic enzymes for many weeks.

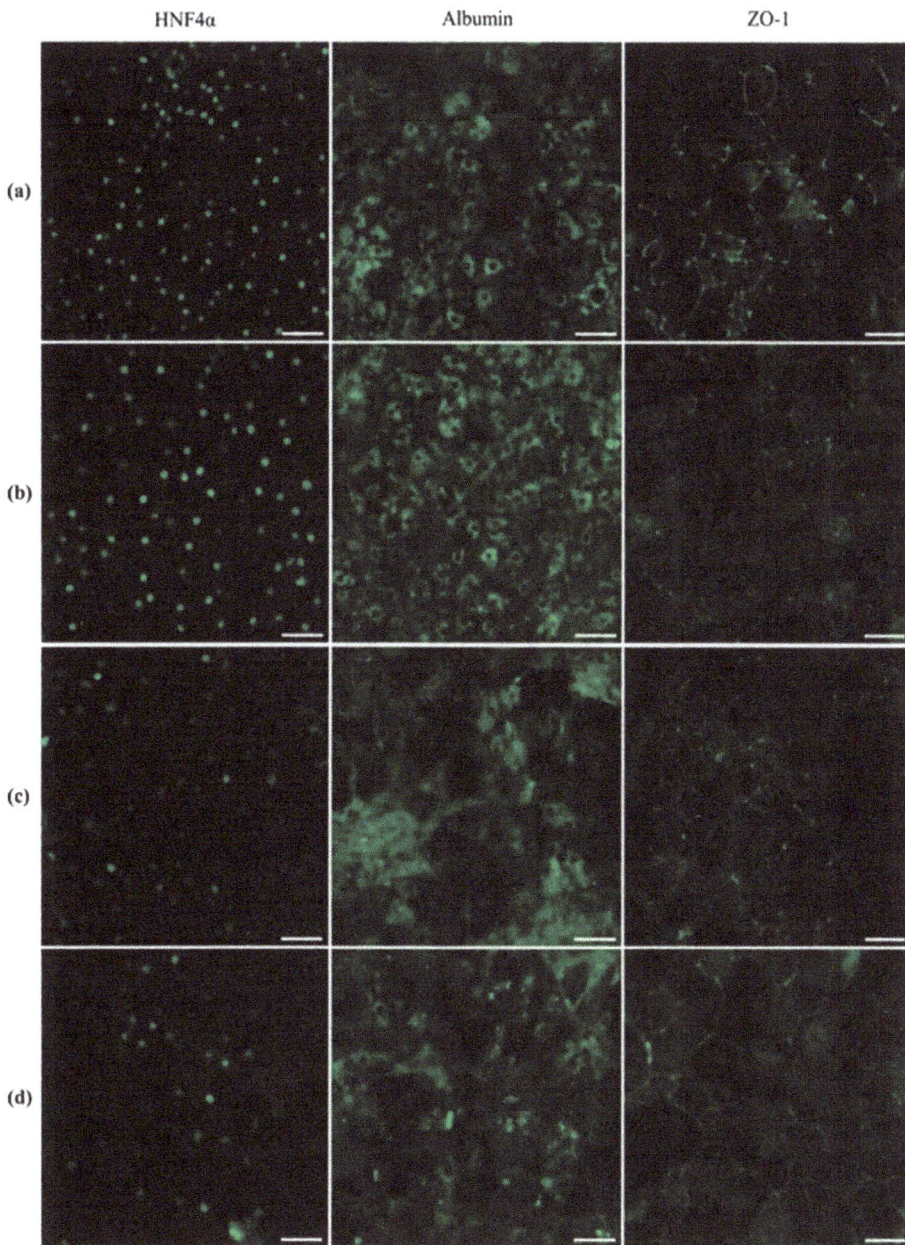

Figure 7. Fluorescence images of printed human-induced pluripotent stem cells (hiPSC)-derived hepatocytes showing hepatocyte marker expression in green: (**a**,**b**) human embryonic stem cells (hESC)-derived hepatocyte-like cells (HLCs) (RC-10): (**a**) Non-printed control; (**b**) printed results; (**c**,**d**) hiPSC-derived HLCs (RCi-22); (**c**) non-printed control; (**d**) printed results (scale bars 50 μm) [81].

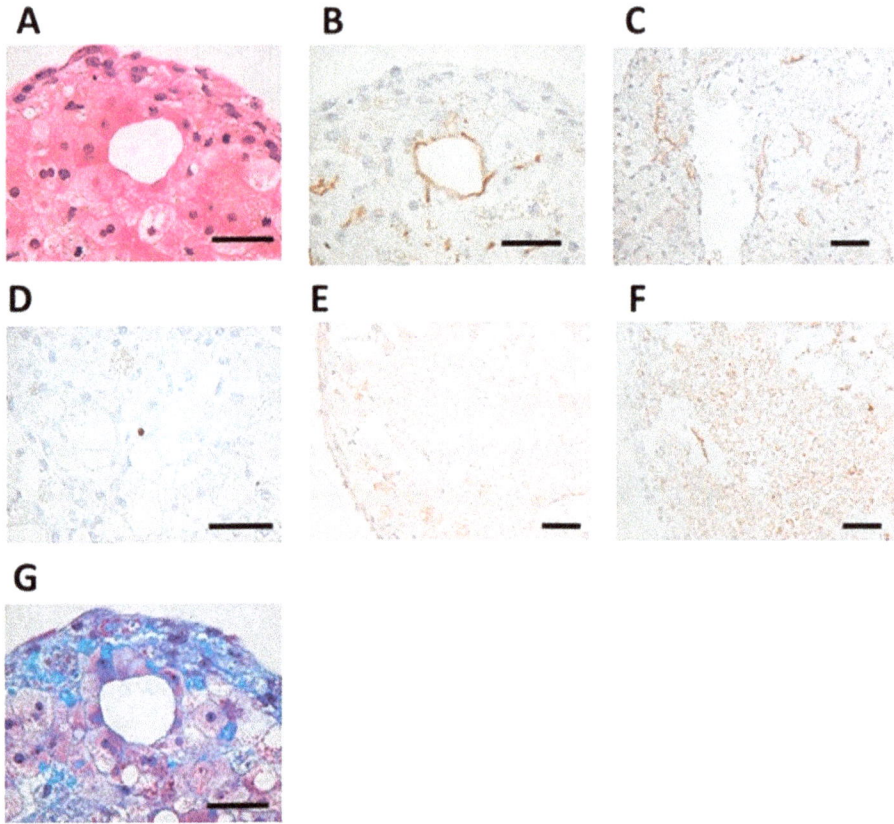

Figure 8. Self-organization in bio-printed human liver tissues. (**A**) Hematoxylin and eosin stain (HE) staining shows structure of bio-printed liver tissue on day 50. (**B**) Immunostaining with the MRP2 antibody detected bile acid transporters (day 50). (**C**) Immunostaining with, cluster of differentiation 31 (CD31) antibody detected blood vessel-like and sinusoid-like structures (day 14). (**D**) Terminal deoxynucleotidyl transferase dUTP nick end labeling (TUNEL) staining detected little apoptosis (day 60). (**E**) Immunostaining with the OAT2/8 antibody detected drug uptake transporters (day 44). (**F**) Immunostaining with MRP2 antibody showed tissue distribution (day 44). (**G**) Masson's trichrome staining shows collagen accumulation (day 50). Black bars represent 50 μm [82].

Tissue engineers have continued to improve the quality of their human liver creations. The creation of living mini-organs is a relatively new area of science with the potential to replace animal models that are not always accurate. Organoid systems are the recently developed 3D bioengineered platforms for studying assays such as drug toxicity testing and metabolic diseases. Organoids are cell-derived in vitro 3D organ models that allow for the study of biological processes and also have important effects for clinical use in an environment that mimics endogenous cell organization and organ structures. These models overcome the major constraints of 2D tissue models and provide prolonged cell viability and functionality [90]. These in vitro culture systems contain a self-renewing stem cell population which differentiates into multiple, organ-specific cell types that exhibit a spatial organization similar to the corresponding organ and are capable of recapitulating some functions of that organ, thus providing a highly physiologically relevant system.

Organoids have been formed via several different methods, e.g., spinner flask cultures [91], utilizing rotating cultures [92], stationary cultures in hanging drops with well-known 96- or 384-well plates [93], and cell growth on non-adherent surfaces [94]. The utilization of engineering tools such as biomaterial scaffolds, microfluidics and bioprinting has enabled greater control over the cellular environment, which has increased the accurate prediction of clinically relevant outcomes and the longevity of liver functions in vitro. For example, Norona et al. [95] fabricated a 3D bioprinted liver tissue housed in a 24-well Transwell (Corning Inc, Corning, NY, USA) that can recapitulate drug-, chemical-, and Transforming growth factor β1 (TGF-β1)-induced fibrogenesis at the cellular, molecular, and histological levels, as demonstrated in Figure 9. Taking into consideration the above characteristics, these bioprinted in vitro tissue models of human liver demonstrate the utility of novel 3D bioprinted tissues to further evaluate compound-induced liver fibrosis in a more defined and systematic way.

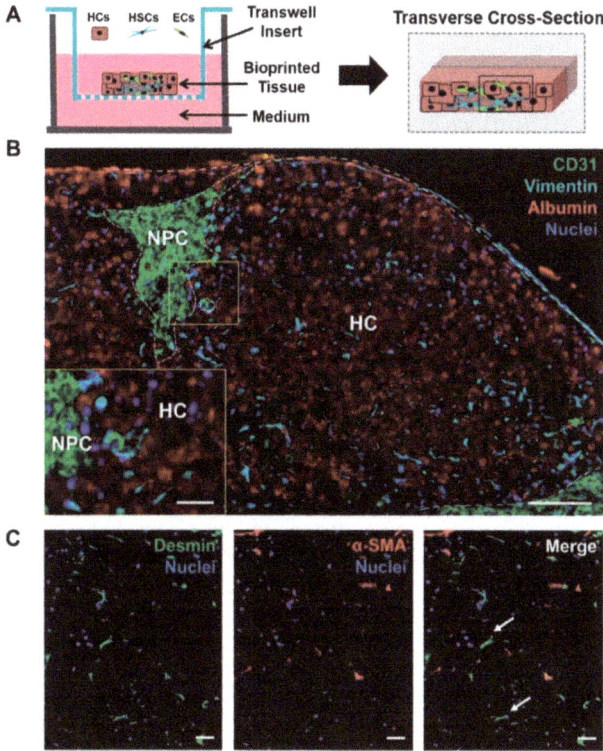

Figure 9. 3D bioprinted tissue exhibits a compartmentalized architecture and maintains hepatic stellate cells in a quiescent-like phenotype. (**A**) Illustration of a transverse cross-section of bioprinted tissue on a transwell insert comprising hepatocytes (HCs) and compartmentalized endothelial cells (ECs) and hepatic stellate cells (HSCs). (**B**) The organization of non-parenchymal cells (NPCs) is depicted with CD31 and vimentin staining to mark ECs and HSCs, respectively. Albumin is used to denote the hepatocellular compartment (HC). Scale bar = 100 μm, inset scale bar = 25 μm. (**C**) HSC activation status was examined using desmin (generic marker) and Alpha-smooth muscle actin (α-SMA) (activation marker). Quiescent HSCs are denoted with white arrows. Scale bar = 50 μm [95].

3.3. Liver-on-Chip Platforms

In contrast to static models, perfusion systems or cell microfluidic platforms can allow for the automated control over several conditions such as culture medium, pH, temperature, fluid pressures, cell shear stress, nutrient supply, and waste removal. Microfluidic systems have been implemented in engineering liver tissues [96]. Significant applications of microfluidics in tissue engineering technology include cell culture and making gradient biomaterials [97]. For these reasons, microfluidic cell platforms are preferable for mimicking the native and dynamic cellular environment compared to static cell culture systems [98]. Moreover, these systems remain precise long term and could provide information on tissue responses to various conditions over time scales that are clinically relevant [99].

The microarchitecture of the liver is crucial to liver function [100]. Hepatocytes interact with mesenchymal cells, stellate cells, Küpffer cells, macrophages, and lymphocytes [101]. A main feature of the liver is the perfusion of fluid. When compared to a conventional cell culture, liver function can be enhanced in a microfluidic chip [102].

Furthermore, some diseases or injury states have also been supported inside a microfluidic chamber for pharmaceutical testing [103,104]. Recently, polydimethylsiloxane (PDMS)-based microfluidic devices have been made obtainable by using multiple chambers to mimic the sinusoidal architecture of the liver. For example, Kang et al. [105] used their system to analyze the viral replication for hepatotropic hepatitis B virus. Moreover, they demonstrated that primary rat hepatocytes maintained normal morphology and produced urea for 30 days when they were cultivated on one side of a transwell membrane, while immortalized bovine aortic endothelial cells were cultivated on the other side of the membrane that was subjected to dual-channel microfluidic perfusion. Another group [106] developed a system to model alcohol injury. Their liver injury-on-chip system was made by two chambers for seeding of hepatocytes and stellate cells, as well as three more chambers for miniature aptamer-modified electrodes to monitor liver cell signaling. This system makes it possible to monitor the paracrine crosstalk between co-cultured cell types communicating via the same signaling.

Additionally, the advantages of perfusion on the functions of liver co-cultures is that perfusion can drive the cells to gradients of oxygen, nutrients, and hormones, which have been shown to lead to liver parenchyma or differential functions in hepatocytes across the length of the sinusoid [107]. Allen et al. [108] fabricated a perfusion bioreactor platform with oxygen gradients that was used to induce an in vivo-like zonal pattern of CYP450s and acetaminophen toxicity in rat hepatocyte cultures. This bioreactor system could provide useful information about the maintenance of liver zonation in order to get deeper insight into the mechanism of metabolism and toxicity.

In contrast to an oxygen gradient, McCarty et al. [109] demonstrated a gradient of exogenous hormone (insulin and glucagon) onto a rat hepatocyte monolayer using a microfluidic device. Utilizing this advanced control system, they demonstrated the in vitro creation of hepatocyte carbohydrate, nitrogen, alcohol degradation, and drug conjugation metabolic zonation. This useful type of system could be essential for the development of in vitro liver disease models.

Only a few reports have been published which combine direct printing techniques with on-chip technologies for the fabrication of organs on chips. Direct printing into a microfluidic chamber to build a liver-on-a-chip platform was also demonstrated by Bhise et al. [110]. Droplets of HepG2 spheroid-Gelatin-methacryloyl (GelMA) mixture were printed on a glass slide within the cell culture chamber of a bioreactor, followed by immediate UV cross linking. The engineered hepatic construct remained functional during the 30-day culture period and showed a drug response similar to published data (Figure 10).

Another printing technique utilizing micro valves integrated with microfluidic chips was studied by Chang et al. [111] in order to fabricate reproducible three-dimensional cell-encapsulated alginate-based, tissue-engineered constructs in chambers for drug screening platforms in planetary environments.

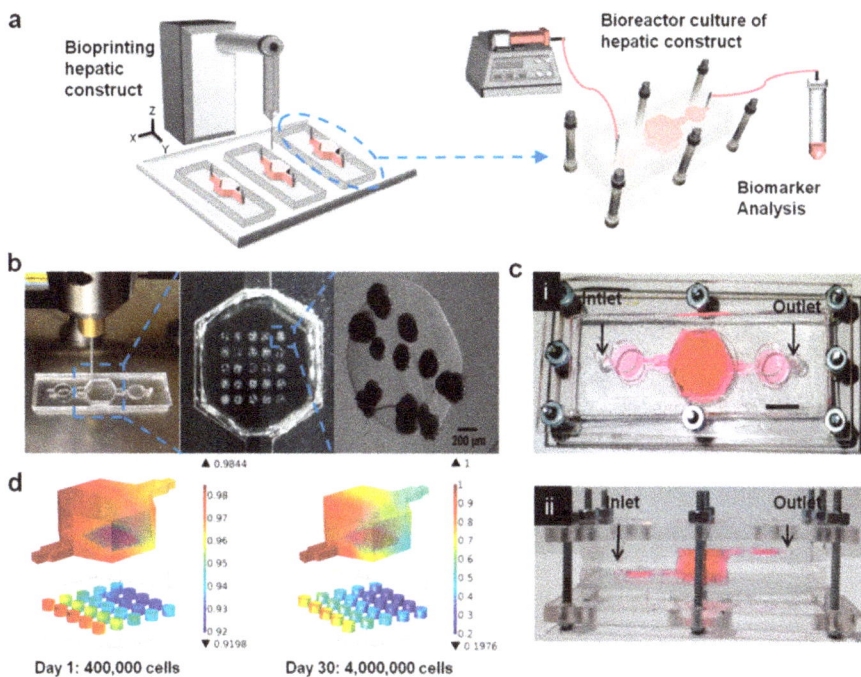

Figure 10. (**a**) Schematic of the hepatic bioreactor culture platform integrated with a bioprinter and biomarker analysis module. (**b**) Bioprinting photocrosslinkable Gelatin-methacryloyl (GelMA) hydrogel-based hepatic construct within the bioreactor as a dot array. (**c**) Top-view (i) and side-view (ii) of the assembled bioreactor with the inlet and outlet fluidic ports as indicated. Scale bar = 1 mm. (**d**) Oxygen concentration gradient in the bioreactor, considering the oxygen uptake of, case A: 400,000 hepatocytes on day one (16,000 cells per dot), and case B: 4000,000 hepatocytes on day 30 (160,000 cells per dot) [110].

Liver platforms are being integrated with different cell lines for liver tissue fabrication. It has been researched that perfused hepatocyte-endothelial co cultures show a greater rate of production of drug metabolites relative to static controls [112]. An interesting in vitro hepatic model was demonstrated by Khetani et al. [60] for drug screening and modeling liver diseases using engineered micropatterned co-cultures of induced pluripotent stem cell-derived human hepatocyte-like cells (iHeps) and 3T3-J2 murine embryonic fibroblasts with a Matrigel. This in vitro model of human liver was maintained for several weeks in culture. Moreover, Cho et al. [113] developed a controlling co-cultured microenvironment to study the heterotypic cell interactions of hepatocytes on a patterned fibroblast layer using microfabricated PDMS stencils. The liver-specific functions of the hepatocytes including intracellular albumin staining and E-cadherin expression were increased as a result of enhanced heterotypic contact in the co culture system. In other similar research, primary human hepatocytes along with human endothelial (EA.hy926), immune (U937) and stellate (LX-2) cells were co-cultured in a microfluidic device. This study described a relevant liver model which was maintained for weeks in order to investigate liver studies and the microfluidic integration technology with other organs [114]. Other approaches to create artificial, three-dimensional hepatic tissue constructs and the regeneration of injured livers reported the co culture systems of hepatic stellate cells (HSCs) [115] and both HSCs and ECs [116]. Several liver platforms have already been fabricated with the aim of the reliable replication of liver physiology and metabolism to benefit the pharmaceutical industry in drug discovery and development. The performance of current liver platforms needs to be improved to further mimic the physiology and function of liver in the body. Future advances in this area could emerge from the combinatory use of existing technologies to move toward a liver model with a more complete functionality.

4. Scaffolds Fabrication Methods

Scaffolds are 3D artificial biostructures which are used in tissue engineering as well-defined matrices for cell adhesion and proliferation. A high porous architecture and a controllable porous size are key parameters for accommodating different types of cells, whereas porosity has a crucial role in attachment and migration of transplanted cells. Depending on the fabrication method and the raw material, the porous size varies between 100 and 500 μm in order to be suitable for applications such as bone regeneration [117], cardiac tissues [118] and cells proliferation [119].

Its biocompatibility, mechanical properties, and chemical properties make the material suitable for medical applications and cell culture. Towards the fabrication of 3D scaffolds, several approaches have been used, such as two-photon polymerization, selective laser sintering, and 3D printing techniques (inkjet and extrusion printing).

4.1. Laser-Based Methods

The main purpose for the fabrication of 3D structures that are aimed to be used as a matrix for the selective placement and growth of cells is the printing of biocompatible polymers for the creation of a 3D shape. Two photon polymerization, a widely used method for developing 3D materials suitable for cell growth and proliferation, is based on the irradiation of a monomer with a laser beam to trigger a cross-linking process by two photon absorption in selected depths [120]. As the desired structure forms by the selective polymerization offered by the laser beam, the non-polymerized monomer is subsequently removed by extensive washing procedures.

The use of lasers for the creation of biopolymer scaffolds enables the easy tuning of the porosity of the final 3D structure by the alteration of the irradiation conditions, as explained by Rekštyte et al. [121]. In the reported study, 3D polymeric porous scaffolds with size porosity of micrometers were obtained with the use of four different combinations of materials and a large variety of fabrication parameters (Figure 11).

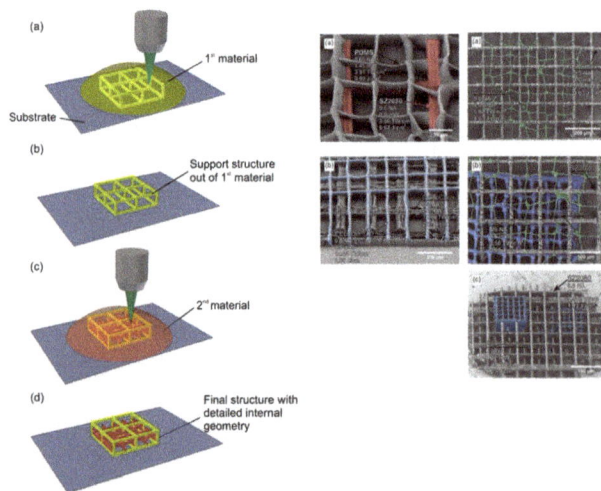

Figure 11. Direct writing laser procedure (**left**). Final structures of fabricated scaffolds consist of two different polymeric materials (**right**) [121].

In addition, Ovsianikov et al. at 2011 [122] created gelatin-based scaffolds with methacrylamide groups for the development of adipose tissue and transplants for plastic surgeries. The results verified the stability of the material and their ability to support ASC adhesion and proliferation from seven to twenty-two days as shown in Figure 12.

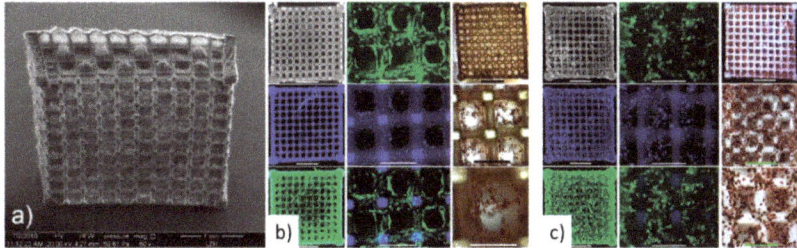

Figure 12. (**a**) SEM image of fabricated gelatin scaffold. (**b**,**c**) Fluorescence microscopy pictures for the 2 photon polymerization scaffold after seven days and 22 days [122].

3D hydrogel scaffolds created by two-photon polymerization (2PP) for the support of Henrietta Lacks (HELA) cells' culture for tissue engineering applications were also reported by Y.C Zheng et al. [123]. The starting material consisted of an aqueous solution of 3,6-bis[2-(1-methyl-pyridinium)vinyl]-9-pentyl-carbazole diiodide (BMVPC), cucurbit [7] uril (CB7), and polyethylene glycol diacrylate (PEGDA) was used as a monomer for 2PP.

Another advantage that laser-based techniques offer is the use of lasers for the creation of the 3D matrix and the selective deposition of cells with high precision. Ovsianikov et al. [120] presented this approach by utilizing lasers to polymerize an acrylated poly(ethyleneglycol) (PEG) monomer for the fabrication of a cell scaffold and the direct laser printing of two different types of cells on the fabricated scaffold (Figure 13).

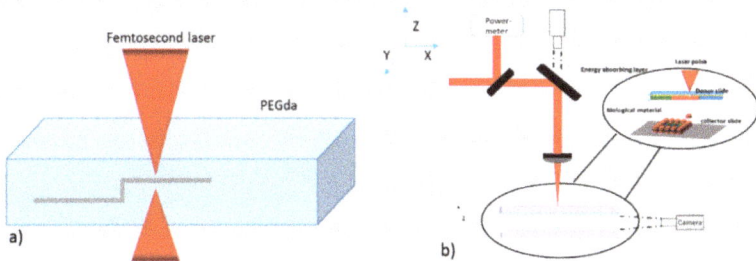

Figure 13. (**a**) Schematic representation of two photon polymerization process of an acrylated poly(ethyleneglycol) (PEG). (**b**) Schematic representation of LIFT technique for the printing of cells on the fabricated scaffold [120].

The final structure of this study had a hexagonal shape with six layers of cylinders along the diameter of the shape (Figure 14). Vascular smooth muscle cells (VSMCs) were laser printed at the outer perimeter of the scaffold, while EC cells were deposited at the inner perimeter (Figure 14).

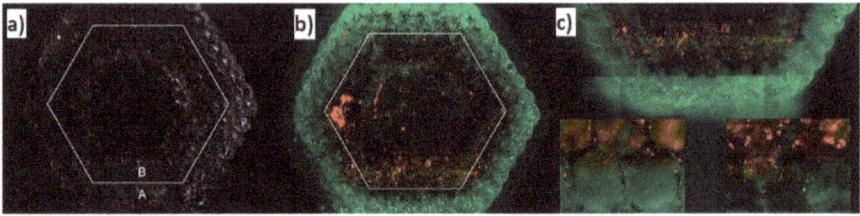

Figure 14. (**a**) SEM images of 2PP fabricated scaffolds. (**b,c**) Fluorescence microscopy images after the deposition of two cell lines in the same scaffold [120].

The combination of the laser-based 2PP technique with the micromolding technique resulted in the accelerated duration of the fabricated scaffolds, according to A. Koroleva et al. [124] (Figure 15). In this structure, human pulmonary microvascular endothelial cells (HPMEC) were cultured for seven days and migrated into the fabricated fibrin gel scaffold (Figure 15).

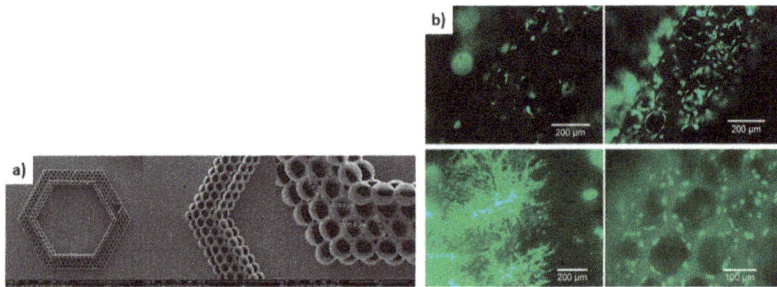

Figure 15. (**a**) Fabricated fibrin gel scaffold with hexagonal shape. (**b**) Cells cultured for seven days [124].

Another laser-based technique used for the fabrication of cell scaffolds is called selective laser sintering [125]. This is a layer-by-layer approach in which a laser beam is used to selectively sinter particles of a polymeric material in order to create layers with specific geometric characteristics [126].

4.2. Inkjet Printing

Inkjet printing is one of the most popular 3D printing techniques for the fabrication of structures of a great variety of materials. This technique enables the printing of picoliter droplets according to a software design in order to create 2D or 3D structures, and it can be either a continuous or a drop-on-demand (DOD) printing approach. The DOD inkjet printing technique has been widely used to create arrays of small liquid biodroplets [127,128]. The mechanism based on a thermal approach or a piezoelectric approach are the two main printing mechanisms with a DOD inkjet printer. The thermal inkjet printer contains of a thermal actuator which heats up the printing head, consequently generating a bubble of gas which, upon expansion, ejects a droplet of liquid to a receiver substrate [129]. On the other hand, the piezoelectric printers consist of a piezoelectric actuator which surrounds the ink chamber. An increase of the voltage across of the piezoelectric actuator initiates the formation of droplets during the flow of the ink [130].

Ink jet printing techniques have been frequently used for the control of cells growth in a matrix by the printing of protein solutions [131], the printing of cells [131], or the printing of the 3D scaffold. In the field of inkjet printing of scaffolds, impressive results were presented by Xu et al. [132] with the printing of a functional 3D scaffold for cardiac tissue application (Figure 16a). Also, Duan et al. who printed valve network (Figure 16b) [54].

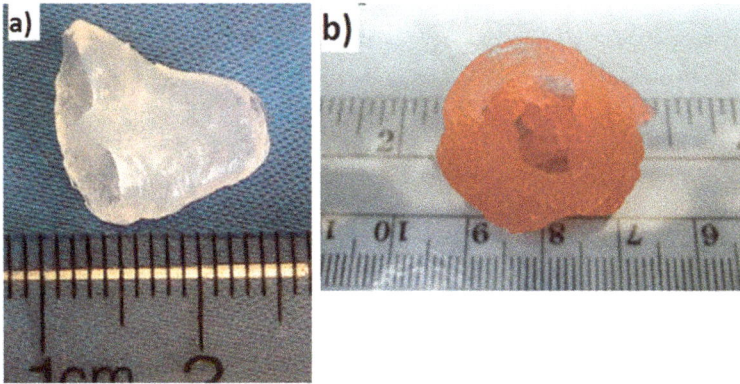

Figure 16. (**a**): Cardiac tissue by Xu et al. [132]. (**b**): Printed valve network by Duan et al. [54].

Moreover, there have been studies that utilize the inkjet printing technique for the creation of a 3D polymer matrix and the precise deposition of cells in a twostep procedure [133].

Even though the inkjet printing technique is an established technique for the printing of materials that can be used as 3D scaffolds for cell cultures, such as biodegradable polymers [134–136] and natural polymers [133,137], most of the reported 3D liver cell cultures by the use of scaffolds are created by more verified techniques in the industry such as microextrusion printing. Microextrusion printing is an additive manufacturing method for creating 3D micro-structures in a layer-by-layer manner. This is enabled by the continuous microprinting of polymeric materials for the creation of individual layers. These types of printers consist of a piston, upon which extraction deposits the biomaterial through a micro-needle. To our knowledge, the only study in which a piezoelectric inkjet printer was used for creating scaffolds for liver cultures was presented by Arai et al. [73]. The novelty of that work was the architecture of the final structure—a sandwich shape of two galactosylated alginate (GA)-gel sheets on the top and bottom and hepatocytes cells in-between the two layers. This design provided the opportunity to regulate the polarity of the hepatocytes.

5. Scaffolds for Liver Tissue Engineering

As mentioned previously, the liver is one of the most important and largest organs in human body, and it plays a significant role in metabolic functions. Many groups have studied the generation of 3D liver structures for potential liver regeneration applications. The use of 3D polymeric structures for generating cell cultures and liver models has facilitated the overcoming of the limitations of 2D cell culture models such as the uncontrollable cells' polarity and non-directed cells' attachment. A wide range of biocompatible polymers has been used as starting materials for building stable matrices for the growth and proliferation of cancer and primary hepatic cell lines. Lewis et al. [138] investigated the creation of a 3D porous gelatin scaffold using a pneumatic extrusion piston-driven EnvisionTEC (GmbH) 3D-Bioplotter.

Six different single layers with different porous sizes were precisely placed in such a way to create two variable geometries (differing in the strength of the connection between the layers), towards the final 3D scaffold. The scaffolds were used as matrix for the cell culture of the differentiated hepatocyte-derived carcinoma cell line (Huh7). After, the fabrication of the optimum scaffold geometry, the viability and the functionality of the seeded Huh7 cells were studied for seven days. Towards this goal, a comparison was made between the two geometries of the 3D scaffolds and the 2D models. The results revealed almost the same viability of the 2D and 3D cell cultures; however, the 3D structures enabled an increase in the hepatic functions of the cells, mainly due to the strong connection between the pores of the structure. These lateral architectural 3D models have been proven suitable to use for studying the specific functions of hepatocytes such as albumin secretion, CYP activity, and bile transport, because they provide an appropriate environment for well-defined cells [138]. Furthermore, a micro-extrusion bioprinter (INKREDIBLE+) was used by Hiller et al. to create a 3D structure by printing a mixed ink consisting of: hydrogels (alginate, gelatin), human extracellular matrix (hECM) and human HepaRG liver cells. The selected cell line was used in thus metabolic study due to its morphology and metabolic characteristics. The aim of this study was to test the metabolic activity and the viability of the cells for structures with variable concentration of hECM. The hECM substance changes the mechanical characteristics of the 3D microstructures. It consists of collagen type I, which improves the properties of the scaffold but, in high concentrations, has a negative effect on cell functionality [80]. Another group (Kim et al.) used alginate and isolated mouse primary hepatocytes to create a 3D bio-printed structure. The process combined the use of a micro-syringe with a three dimensional motion stage to create the final 3D structure in a layer-by-layer manner. The final shape of the structure was adjusted by scanning parameters such as velocity and pressure. The main goal of this study was to create hepatocyte cell culture networks which demonstrated a high viability of the cells after 14 days with good hepatic functionality [75]. The same year, Kang et al. [139] used mouse-induced hepatocyte-like cells (miHeps) by pluripotent stem cells (PSC) for the development of a 3D structure made by miHeps and aginate using extrusion printing. Five layers of cells and alginate printing solution built the final 3D cell culture which was placed in a mouse in vivo. The implant was examined 14 and 28 days after the surgery, presenting results that indicated that the in vivo transplanted scaffold was more functional than the in vitro model. Lee et al. [76] created 3D scaffolds with improved mechanical properties for a 3D hepatocytes cell culture environment. In this study, the scaffold was made by polycaprolactone (PCL) as a starting material by using a homemade printer with multiple deposition heads-multihead tissue/organ building (MtoBS system). The layer-by-layer printed structure consisted of PCL and hydrogel layers with a mixture of collagen and three different cell lines—HCs, HUVECs, and HLFs. The final structures had the ability to support multiple functional cell lines which could maintain their hepatic functions for 10 days. Jeon et al. fabricated 3D alginate scaffolds with cancer hepatic cells (HepG2) using a micro extrusion printer and tested the proliferation and the viability of hepatocytes on the 3D structure for three weeks. The histology and immunohistochemistry of the final cultures were investigated along with their ability to support operational cells [140]. Gong et al. designed and fabricated well defined 3D chitosan–gelatin (C/G) scaffolds which consisted of polymeric channels and pores using both an indirect method called the solid freeform fabrication (SFF) and freeze drying methods. Two thermoplastic materials were used for forming the mold of the scaffold. Chitosan and gelatin were used as matrix materials, and the final structure was initiated by the freeze-drying process, which also led to the creation of micro pores. The tuning of the freeze-drying parameters resulted in the optimum shape and morphology of the scaffold. The functionality of the developed scaffold structure was tested for the culturing of HepG2 cell line [141].

6. Conclusions

In summary, 3D bioprinting technology enables the fabrication of biomimetic tissues and implants with the use of biomaterials, growth factors, and living cells, which can either be printed in a specific pattern for the development of the final tissue structure or, in many cases, can be printed on an already existing 3D matrix (scaffold). Furthermore, in the field of tissue engineering, an bioartificial liver is considered one of the most promising tools as a therapeutic method for severe liver diseases and, in the field of regenerative medicine, for drug testing. The most commonly used direct writing techniques for the printing of cells are laser-based techniques, inkjet printing, and microextrusion printing. Those techniques are mainly chosen because they can easily adapt to the cultivation environment, can create high-resolution cell structures, and, in many cases, can also be used to create 3D scaffolds for the cell growth. In the field of liver tissue engineering, a lot of work has been done with the use of the above-mentioned techniques towards; however, greater effort is required to solve problems

Bioengineering **2019**, *6*, 95

encountered in the inability to replicate the actual 3D living liver tissue environment. Finally, the combination of a 3D bioprinting technique with a microfluidic control can be a promising method for controlled drug delivery systems and for future regenerative medicine.

Author Contributions: Individual contributions of authors are as follows: Writing—original draft preparation, C.K., V.L.; Writing—review and editing, M.C., I.Z.; Supervision, I.Z.

Funding: This research was funded by State Scholarships Foundation: 'Reinforcement of Postdoctoral Researchers'.

Acknowledgments: The work was supported by the IKY scholarships program, which is co-financed by the European Union (European Social Fund-ESF) and Greek national funds through the action entitled 'Reinforcement of Postdoctoral Researchers', in the framework of the Operational Programme 'Human Resources Development Program, Education and Lifelong Learning' of the National Strategic Reference Framework (NSRF) 2014–2020 and by ESPA 2014–2020 program through the action entitled Synergy of ELI-LASERLAB Europe, HIPER&IPERION-CH.gr in the framework of the NTUA's Participation in the project HELLAS-CH (MIS 5002735).

Conflicts of Interest: The authors declare no conflict of interest.

References

1. Kruth, J. Material Incress Manufacturing by Rapid Prototyping Techniques. *CIRP Ann.* **1991**, *40*, 603–614. [CrossRef]

2. Heller, T.B.; Hill, R.M.; Saggal, A.F. Method of and Apparatus for Forming a Solid Three-Dimensional Article from a Liquid Medium. U.S. Patent 5,071,337, 10 December 1991.

3. Hull, C.W. Apparatus for Production of Three-Dimensional Objects by Stereolithography. U.S. Patent 4,575,330, 11 March 1986.

4. Nakamura, M.; Iwanaga, S.; Henmi, C.; Arai, K.; Nishiyama, Y. Biomatrices and biomaterials for future developments of bioprinting and biofabrication. *Biofabrication* **2010**, *2*, 014110. [CrossRef] [PubMed]

5. Murphy, S.V.; Atala, A. 3D bioprinting of tissues and organs. *Nat. Biotechnol.* **2014**, *32*, 773–785. [CrossRef] [PubMed]

6. Munaz, A.; Vadivelu, R.K.; John, J.S.; Barton, M.; Kamble, H.; Nguyen, N.T. Three-dimensional printing of biological matters. *J. Sci. Adv. Mater. Devices* **2016**, *1*, 1–17. [CrossRef]

7. Derakhshanfar, S.; Mbeleck, R.K.; Xu, K.; Zhang, X.; Zhong, W.; Xing, M. 3D bioprinting for biomedical devices and tissue engineering: A review of recent trends and advances. *Bioact. Mater.* **2018**, *3*, 144–156. [CrossRef] [PubMed]

8. Bertassoni, L.E.; Cecconi, M.; Manoharan, V.; Nikkhah, M.; Hjortnaes, J.; Cristino, A.L.; Barabaschi, G.; Demarchi, D.; Dokmeci, M.R.; Yang, Y.; et al. Hydrogel bioprinted microchannel networks for vascularization of tissue engineering constructs. *Lab Chip* **2014**, *14*, 2202–2211. [CrossRef] [PubMed]

9. Moya, M.L.; Hsu, Y.H.; Lee, A.P.; Hughes, C.C.; George, S.C. In Vitro Perfused Human Capillary Networks. *Tissue Eng. Part C Methods* **2013**, *19*, 730–737. [CrossRef] [PubMed]

10. Li, X.; Chen, Y.; Li, P.C.H. A simple and fast microfluidic approach of same-single-cell analysis (SASCA) for the study of multidrug resistance modulation in cancer cells. *Lab Chip* **2011**, *11*, 1378–1384. [CrossRef]

11. Li, X.J.; Nie, Z.H.; Cheng, C.M.; Goodale, A.B.; Whitesides, G.M. Paper-based electrochemical ELISA. *Proc. Micro Total Anal. Syst.* **2010**, *14*, 1487–1489.

12. Salieb-Beugelaar, G.B.; Simone, G.; Arora, A.; Philippi, A.; Manz, A. Latest Developments in Microfluidic Cell Biology and Analysis Systems. *Anal. Chem.* **2010**, *82*, 4848–4864. [CrossRef]

13. Jang, K.; Sato, K.; Igawa, K.; Chung, U.I.; Kitamori, T. Development of an osteoblast-based 3D continuous-perfusion microfluidic system for drug screening. *Anal. Bioanal. Chem.* **2008**, *390*, 825–832. [CrossRef] [PubMed]

14. Chang, C.C.; Boland, E.D.; Williams, S.K.; Hoying, J.B. Direct-write bioprinting three-dimensional biohybrid systems for future regenerative therapies. *J. Biomed. Mater. Res. Part B Appl. Biomater.* **2011**, *98*, 160–170. [CrossRef] [PubMed]

15. Kurobe, H.; Maxfield, M.W.; Breuer, C.K.; Shinoka, T. Concise Review: Tissue-Engineered Vascular Grafts for Cardiac Surgery: Past, Present, and Future. *Stem Cells Transl. Med.* **2012**, *1*, 566–571. [CrossRef] [PubMed]

16. Freeman, R.B., Steffick, D.E., Jr.; Guidinger, M.K.; Farmer, D.G.; Berg, C.L.; Merion, R.M. Liver and Intestine Transplantation in the United States 1998–2006. *Am. J. Transplant.* **2008**, *8*, 958–976. [CrossRef] [PubMed]

17. Powers, M.J.; Janigian, D.M.; Wack, K.E.; Baker, C.S.; Stolz, D.B.; Griffith, L.G. Functional Behavior of Primary Rat Liver Cells in a Three-Dimensional Perfused Microarray Bioreactor. *Tissue Eng.* **2002**, *8*, 499–513. [CrossRef] [PubMed]

18. Allen, J.W.; Bhatia, S.N. Improving the next generation of bioartificial liver devices. *Semin. Cell Dev. Biol.* **2002**, *13*, 447–454. [CrossRef] [PubMed]

19. Bao, J.; Fisher, J.; Nyberg, S.L. Liver Regeneration and Tissue Engineering. In *Tissue Engineering in Regenerative Medicine*; Humana Press: New York, NY, USA, 2011; pp. 315–332.

20. Melchels, F.P.W.; Domingos, M.A.N.; Klein, T.J.; Malda, J.; Bartolo, P.J.; Hutmacher, D.W. Additive Manufacturing of Tissues and Organs. *Prog. Polym. Sci.* **2012**, *37*, 1079–1104. [CrossRef]

21. Jose, R.R.; Rodriguez, M.J.; Dixon, T.A.; Omenetto, F.G.; Kaplan, D.L. Evolution of Bioinks and Additive Manufacturing Technologies for 3D Bioprinting. *ACS Biomater. Sci. Eng.* **2016**, *2*, 1662–1678. [CrossRef]

22. Gross, B.C.; Erkal, J.L.; Lockwood, S.Y.; Chen, C.; Spence, D.M. Evaluation of 3D Printing and Its Potential Impact on Biotechnology and the Chemical Sciences. *Anal. Chem.* **2014**, *86*, 3240–3253. [CrossRef] [PubMed]

23. Choi, J.Y.; Das, S.; Theodore, N.D. Advances in 2D/3D Printing of Functional Nanomaterials and Their Applications. *ECS J. Solid State Sci. Technol.* **2015**, *4*, 3001–3009. [CrossRef]

24. Wang, X.; Yan, Y.; Zhang, R. Rapid prototyping as a tool for manufacturing bioartificial livers. *Trends Biotechnol.* **2007**, *25*, 505–513. [CrossRef] [PubMed]

25. Derby, B. Printing and prototyping of tissues and scaffolds. *Science* **2012**, *338*, 921–926. [CrossRef] [PubMed]

26. Choi, Y.J.; Yi, H.G.; Kim, S.W.; Cho, D.W. 3D Cell Printed Tissue Analogues: A New Platform for Theranostics. *Theranostics* **2017**, *7*, 3118–3137. [CrossRef] [PubMed]

27. Iwami, K.; Noda, T.; Ishida, K.; Morishima, K.; Nakamura, M.; Umeda, N. Bio rapid prototyping by extruding/aspirating/refilling thermoreversible hydrogel. *Biofabrication* **2010**, *2*, 014108. [CrossRef] [PubMed]

28. Xu, T.; Zhao, W.; Zhu, J.M.; Albanna, M.Z.; Yoo, J.J.; Atala, A. Complex heterogeneous tissue constructs containing multiple cell types prepared by inkjet printing technology. *Biomaterials* **2013**, *34*, 130–139. [CrossRef]

29. Guillemot, F.; Souquet, A.; Catros, S.; Guillotin, B.; Lopez, J.; Faucon, M.; Pippenger, B.; Bareille, R.; Remy, M.; Bellance, S.; et al. High-throughput laser printing of cells and biomaterials for tissue engineering. *Acta Biomater.* **2010**, *6*, 2494–2500. [CrossRef]

30. Guillotin, B.; Souquet, A.; Catros, S.; Duocastella, M.; Pippenger, B.; Bellance, S.; Bareille, R.; Remy, M.; Bordenave, L.; Amedee, J.; et al. Laser assisted bioprinting of engineered tissue with high cell density and microscale organization. *Biomaterials* **2010**, *31*, 7250–7256. [CrossRef]

31. Bohandy, J.; Kim, B.F.; Adrian, F.J. Metal deposition from a supported metal film using an excimer laser. *J. Appl. Phys.* **1986**, *60*, 1538–1539. [CrossRef]

32. Guillemot, F.; Souquet, A.; Catros, S.; Guillotin, B. Laser-assisted cell printing: Principle, physical parameters versus cell fate and perspectives in tissue engineering. *Nanomedicine* **2010**, *5*, 507–515. [CrossRef] [PubMed]

33. Zhang, X.; Zhang, Y. Tissue Engineering Applications of Three-Dimensional Bioprinting. *Cell Biophys.* **2015**, *72*, 777–782. [CrossRef] [PubMed]

34. Karaiskou, A.; Zergioti, I.; Fotakis, C.; Kapsetaki, M.; Kafetzopoulos, D. Microfabrication of biomaterials by the sub-ps laser-induced forward transfer process. *Appl. Surf. Sci.* **2003**, *208*, 245–249. [CrossRef]

35. Othon, C.M.; Wu, X.; Anders, J.J.; Ringeisen, B.R. Single-cell printing to form three-dimensional lines of olfactory ensheathing cells. *Biomed. Mater.* **2008**, *3*, 034101. [CrossRef] [PubMed]

36. Gruene, M.; Deiwick, A.; Koch, L.; Schlie, S.; Unger, C.; Hofmann, N.; Bernemann, I.; Glasmacher, B.; Chichkov, B. Laser Printing of Stem Cells for Biofabrication of Scaffold-Free Autologous Grafts. *Tissue Eng. Part C Methods* **2011**, *17*, 79–87. [CrossRef] [PubMed]

37. Koch, L.; Deiwick, A.; Schlie, S.; Michael, S.; Gruene, M.; Coger, V.; Zychlinski, D.; Schambach, A.; Reimers, K.; Vogt, P.M.; et al. Skin tissue generation by laser cell printing. *Biotechnol. Bioeng.* **2012**, *109*, 1855–1863. [CrossRef] [PubMed]

38. Koch, L.; Kuhn, S.; Sorg, H.; Gruene, M.; Schlie, S.; Gaebel, R.; Polchow, B.; Reimers, K.; Stoelting, S.; Ma, N.; et al. Laser Printing of Skin Cells and Human Stem Cells. *Tissue Eng. Part C Methods* **2010**, *16*, 847–854. [CrossRef] [PubMed]

39. Catros, S.; Guillotin, B.; Bacakova, M.; Fricain, J.C.; Guillemot, F. Effect of laser energy, substrate film thickness and bioink viscosity on viability of endothelial cells printed by Laser-Assisted Bioprinting. *Appl. Surf. Sci.* **2011**, *257*, 5142–5147. [CrossRef]

40. Keriquel, V.; Guillemot, F.; Arnault, I.; Guillotin, B.; Miraux, S.; Amedee, J.; Fricain, J.C.; Catros, S. In vivobioprinting for computer- and robotic-assisted medical intervention: Preliminary study in mice. *Biofabrication* **2010**, *2*, 014101. [CrossRef] [PubMed]

41. Gudapati, H.; Dey, M.; Ozbolat, I. A comprehensive review on droplet-based bioprinting: Past, present and future. *Biomaterials* **2016**, *102*, 20–42. [CrossRef]

42. Hansen, C.J.; Saksena, R.; Kolesky, D.B.; Vericella, J.J.; Kranz, S.J.; Muldowney, G.P.; Christensen, K.T.; Lewis, J.A. High-Throughput Printing via Microvascular Multinozzle Arrays. *Adv. Mater.* **2013**, *25*, 96–102. [CrossRef] [PubMed]

43. Kim, J.D.; Choi, J.S.; Kim, B.S.; Choi, Y.C.; Cho, Y.W. Piezoelectric inkjet printing of polymers: Stem cell patterning on polymer substrates. *Polymer* **2010**, *51*, 2147–2154. [CrossRef]

44. Xu, T.; Jin, J.; Gregory, C.; Hickman, J.J.; Boland, T. Inkjet printing of viable mammalian cells. *Biomaterials* **2005**, *26*, 93–99. [CrossRef] [PubMed]

45. Skardal, A.; Zhang, J.; Prestwich, G.D. Bioprinting vessel-like constructs using hyaluronan hydrogels crosslinked with tetrahedral polyethylene glycol tetracrylates. *Biomaterials* **2010**, *31*, 6173–6181. [CrossRef] [PubMed]

46. Phillippi, J.A.; Miller, E.; Weiss, L.; Huard, J.; Waggoner, A.; Campbell, P. Microenvironments Engineered by Inkjet Bioprinting Spatially Direct Adult Stem Cells Toward Muscle- and Bone-Like Subpopulations. *Stem Cells* **2008**, *26*, 127–134. [CrossRef] [PubMed]

47. Skardal, A.; Mack, D.; Kapetanovic, E.; Atala, A.; Jackson, J.D.; Yoo, J.; Soker, S. Bioprinted Amniotic Fluid-Derived Stem Cells Accelerate Healing of Large Skin Wounds. *Stem Cells Transl. Med.* **2012**, *1*, 792–802. [CrossRef] [PubMed]

48. Cui, X.; Breitenkamp, K.; Finn, M.; Lotz, M.; D'Lima, D.D. Direct Human Cartilage Repair Using Three-Dimensional Bioprinting Technology. *Tissue Eng. Part A* **2012**, *18*, 1304–1312. [CrossRef] [PubMed]

49. Marga, F.; Jakab, K.; Khatiwala, C.; Shephard, B.; Dorfman, S.; Forgacs, G. Organ printing: A novel tissue engineering paradigm. In Proceedings of the 5th European Conference of the International Federation for Medical and Biological Engineering, Budapest, Hungary, 14–18 September 2011; pp. 27–30.

50. Marga, F.; Jakab, K.; Khatiwala, C.; Shepherd, B.; Dorfman, S.; Hubbard, B.; Colbert, S.; Gabor, F. Toward engineering functional organ modules by additive manufacturing. *Biofabrication* **2012**, *4*, 022001. [CrossRef] [PubMed]

51. Mironov, V.; Kasyanov, V.; Markwald, R.R. Organ printing: From bioprinter to organ biofabrication line. *Curr. Opin. Biotechnol.* **2011**, *22*, 667–673. [CrossRef] [PubMed]

52. Kolesky, D.B.; Homan, K.A.; Skylar-Scott, M.A.; Lewis, J.A. Three-dimensional bioprinting of thick vascularized tissues. *Proc. Natl. Acad. Sci. USA* **2016**, *113*, 3179–3184. [CrossRef] [PubMed]

53. Chang, R.; Nam, J.; Sun, W. Effects of dispensing pressure and nozzle diameter on cell survival from solid freeform fabrication-based direct cell writing. *Tissue Eng. Part A* **2008**, *14*, 41–48. [CrossRef] [PubMed]

54. Duan, B.; Hockaday, L.A.; Kang, K.H.; Butcher, J.T. 3D bioprinting of heterogeneous aortic valve conduits with alginate/gelatin hydrogels. *J. Biomed. Mater. Res. A* **2013**, *101*, 1255–1264. [CrossRef]

55. Chang, R.; Nam, J.; Sun, W. Direct Cell Writing of 3D Microorgan for In Vitro Pharmacokinetic Model. *Tissue Eng. Part C Methods* **2008**, *14*, 157–166. [CrossRef] [PubMed]

56. Widmaier, E.P.; Raff, H.; Strang, K.T. *Vander's Human Physiology*; McGraw-Hill Education: Boston, NY, USA, 2010.

57. Betts, J.G.; Desaix, P.; Johnson, J.E.; Korol, O.; Kruse, D.; Poe, B.; Womble, M.D. *Anatomy & Physiology*; Rice University: Houston, TX, USA, 2013.

58. Eguchi, S.; Chen, S.C.; Rozga, J.; Demetriou, A.A. Tissue Engineering: Liver. In *Yearbook of Cell and Tissue Transplantation 1996–1997*; Kluwer Academic Publishers: Dordrecht, The Netherlands, 1996; pp. 247–252. [CrossRef]

59. Tayyeb, A.; Azam, F.; Nisar, R.; Nawaz, R.; Qaisar, U.; Ali, G. Regenerative Medicine in Liver Cirrhosis: Promises and Pitfalls. In *Liver Cirrhosis—Update and Current Challenges*; BoD Publisher: Norderstedt, Germany, 2017. [CrossRef]

60. Khetani, S.R.; Bhatia, S.N. Microscale culture of human liver cells for drug development. *Nat. Biotechnol.* **2008**, *26*, 120–126. [CrossRef] [PubMed]

61. Hewitt, N.J.; Gómez Lechón, M.J.; Houston, J.B.; Hallifax, D.; Brown, H.S.; Maurel, P.; Kenna, J.G.; Gustavsson, L.; Lohmann, C.; Skonberg, C.; et al. Primary hepatocytes: Current understanding of the regulation of metabolic enzymes and transporter proteins, and pharmaceutical practice for the use of hepatocytes in metabolism, enzyme induction, transporter, clearance, and hepatotoxicity studies. *Drug Metab. Rev.* **2007**, *39*, 159–234. [CrossRef] [PubMed]

62. Griffith, L.G.; Wu, B.; Cima, M.J.; Powers, M.J.; Chaignaud, B.; Vacanti, J.P. In vitro organogenesis of liver tissue. *Ann. N. Y. Acad. Sci.* **1997**, *831*, 382–397. [CrossRef] [PubMed]

63. Takayama, K.; Kawabata, K.; Nagamoto, Y.; Kishimoto, K.; Tashiro, K.; Sakurai, F.; Tachibana, M.; Kanda, K.; Hayakawa, T.; Furue, M.K.; et al. 3D spheroid culture of hESC/hiPSC-derived hepatocyte-like cells for drug toxicity testing. *Biomaterials* **2013**, *34*, 1781–1789. [CrossRef] [PubMed]

64. Gaskell, H.; Sharma, P.; Colley, H.E.; Murdoch, C.; Williams, D.P.; Webb, S.D. Characterization of a functional C3A liver spheroid model. *Toxicol. Res. (Camb.)* **2016**, *5*, 1053–1065. [CrossRef]

65. Ho, C.T.; Lin, R.Z.; Chen, R.J.; Chin, C.K.; Gong, S.E.; Chang, H.Y.; Peng, H.L.; Hsu, L.; Yew, T.R.; Chang, S.F.; et al. Liver-cell patterning Lab Chip: Mimicking the morphology of liver lobule tissue. *Lab Chip* **2013**, *13*, 3578–3587. [CrossRef]

66. Berger, D.R.; Ware, B.R.; Davidson, M.D.; Allsup, S.R.; Khetani, S.R. Enhancing the functional maturity of iPSC-derived human hepatocytes via controlled presentation of cell-cell interactions in vitro. *Hepatology* **2014**, *1*, 1370–1381. [CrossRef]

67. Taub, R. Liver regeneration: From myth to mechanism. *Nat. Rev. Mol. Cell Biol.* **2004**, *5*, 836–847. [CrossRef]

68. Wang, X.; Yan, Y.; Pan, Y.; Xiong, Z.; Liu, H.; Cheng, J.; Liu, F.; Lin, F.; Wu, R.; Zhang, R.; et al. Generation of three-dimensional hepatocyte/gelatin structures with rapid prototyping system. *Tissue Eng.* **2006**, *12*, 83–90. [CrossRef]

69. Robbins, J.B.; Gorgen, V.; Min, P.; Shepherd, B.R.; Presnell, S.C. A novel in vitro three dimensional bioprinted liver tissue system for drug development. *FASEB J.* **2013**, *27*, 812–872.

70. Zein, N.N.; Hanouneh, I.A.; Bishop, P.D.; Samaan, M.; Eghtesad, B.; Quintini, C.; Miller, C.; Yerian, L.; Klatte, R. Three-dimensional print of a liver for preoperative planning in living donor liver transplantation. *Liver Transpl.* **2013**, *19*, 1304–1310. [CrossRef] [PubMed]

71. Nguyen, D.G.; Funk, J.; Robbins, J.B.; Crogan-Grundy, C.; Presnell, S.C.; Singer, T.; Roth, A.B. Bioprinted 3D primary liver tissues allow assessment of organ-level response to clinical drug induced toxicity in vitro. *PLoS ONE* **2016**, *11*, 0158674. [CrossRef] [PubMed]

72. Leva, V.; Chatzipetrou, M.; Alexopoulos, L.; Tzeranis, D.S.; Zergioti, I. Direct Laser Printing of Liver Cells on Porous Collagen Scaffolds. *JLMN J. Laser Micro Nanoeng.* **2018**, *13*, 234–237. [CrossRef]

73. Arai, K.; Yoshida, T.; Okabe, M.; Goto, M.; Mir, T.A.; Soko, C.; Tsukamoto, Y.; Akaike, T.; Nikaido, T.; Zhou, K.; et al. Fabrication of 3D-culture platform with sandwich architecture for preserving liver-specific functions of hepatocytes using 3D bioprinter. *J. Biomed. Mater. Res. Part A* **2017**, *105*, 1583–1592. [CrossRef] [PubMed]

74. Matsusaki, M.; Sakaue, K.; Kadowaki, K.; Akashi, M. Three dimensional human tissue chips fabricated by rapid and automatic inkjet cell printing. *Adv. Healthc. Mater.* **2013**, *2*, 534–539. [CrossRef] [PubMed]

75. Kim, Y.; Kang, K.; Jeong, J.; Paik, S.S.; Kim, J.S.; Park, S.A.; Kim, W.D.; Park, J.; Choi, D. Three-dimensional (3D) printing of mouse primary hepatocytes to generate 3D hepatic structure. *Ann. Surg. Treat. Res.* **2017**, *92*, 67–72. [CrossRef] [PubMed]

76. Lee, J.W.; Choi, Y.J.; Yong, W.J.; Pati, F.; Shim, J.H.; Kang, K.S.; Kang, I.H.; Park, J.; Cho, D.W. Development of a 3D cell printed construct considering angiogenesis for liver tissue engineering. *Biofabrication* **2016**, *8*, 015007. [CrossRef]

77. Skardal, A.; Devarasetty, M.; Kang, H.W.; Mead, I.; Bishop, C.; Shupe, T.; Lee, S.J.; Jackson, J.; Yoo, J.; Soker, S. A hydrogel bioink toolkit for mimicking native tissue biochemical and mechanical properties in bioprinted tissue constructs. *Acta Biomater.* **2015**, *25*, 24–34. [CrossRef]

78. Mazzocchi, A.; Devarasetty, M.; Huntwork, R.C.; Soker, S.; Skardal, A. Optimization of collagen type I-hyaluronan hybrid bioink for 3D bioprinted liver microenvironments. *Biofabrication* **2018**, *11*, 015003. [CrossRef]

79. Lee, H.; Han, W.; Kim, H.; Ha, D.H.; Jang, J.; Kim, B.S.; Cho, D.W. Development of Liver Decellularized Extracellular Matrix Bioink for Three-Dimensional Cell Printing-Based Liver Tissue Engineering. *Biomacromolecules* **2017**, *18*, 1229–1237. [CrossRef] [PubMed]

80. Hiller, T.; Berg, J.; Elomaa, L.; Röhrs, V.; Ullah, I.; Schaar, K.; Dietrich, A.C.; Al-Zeer, M.A.; Kurtz, A.; Hocke, A.C.; et al. Generation of a 3D Liver Model Comprising Human Extracellular Matrix in an Alginate/Gelatin-Based Bioink by Extrusion Bioprinting for Infection and Transduction Studies. *Int. J. Mol. Sci.* **2018**, *19*, 3129. [CrossRef] [PubMed]

81. Faulkner-Jones, A.; Greenhough, S.; King, J.A.; Gardner, J.; Courtney, A.; Shu, W. Development of a valve-based cell printer for the formation of human embryonic stem cell spheroid aggregates. *Biofabrication* **2013**, *5*, 015013. [CrossRef] [PubMed]

82. Faulkner-Jones, A.; Fyfe, C.; Cornelissen, D.J.; Gardner, J.; King, J.; Courtney, A.; Shu, W. Bioprinting of human pluripotent stem cells and their directed differentiation into hepatocyte-like cells for the generation of mini-livers in 3D. *Biofabrication* **2015**, *7*, 044102. [CrossRef] [PubMed]

83. Lei, M.; Wang, X. Biodegradable Polymers and Stem Cells for Bioprinting. *Molecules* **2016**, *21*, 539. [CrossRef] [PubMed]

84. Kim, Y.; Kang, K.; Yoon, S.; Kim, J.S.; Park, S.A.; Kim, W.D.; Lee, S.B.; Ryu, K.Y.; Jeong, J.; Choi, D. Prolongation of liver-specific function for primary hepatocytes maintenance in 3D printed architectures. *Organogenesis* **2018**, *14*, 1–12. [CrossRef] [PubMed]

85. Orlando, G.; Baptista, P.; Birchall, M.; Coppi, P.D.; Farney, A.; Guimaraes-Souza, N.K.; Opara, E.; Rogers, J.; Seliktar, D.; Shapira-Schweitzer, K.; et al. Regenerative medicine as applied to solid organ transplantation: Current status and future challenges. *Transpl. Int.* **2011**, *24*, 223–232. [CrossRef] [PubMed]

86. Chaudhari, P.; Tian, L.; Deshmukh, A.; Jang, Y.-Y. Expression kinetics of hepatic progenitor markers in cellular models of human liver development recapitulating hepatocyte and biliary cell fate commitment. *Exp. Biol. Med.* **2016**, *241*, 1653–1662. [CrossRef]

87. Maza, G.; Rombouts, K.; Hall, A.R.; Urbani, L.; Luong, T.V.; Al-Akkad, W.; Longato, L.; Brown, D.; Maghsoudlou, P.; Dhillon, A.P.; et al. Decellularized human liver as a natural 3D-scaffold for liver bioengineering and transplantation. *Sci. Rep.* **2015**, *5*, 13079. [CrossRef]

88. Hussein, K.H.; Park, K.M.; Ghim, J.H.; Yang, S.R.; Woo, H.M. Three dimensional culture of HepG2 liver cells on a rat decellularized liver matrix for pharmacological studies. *J. Biomed. Mater. Res. Part B Appl. Biomater.* **2016**, *104*, 263–273. [CrossRef]

89. Kizawa, H.; Nagao, E.; Shimamura, M.; Zhang, G.; Torii, H. Scaffold-free 3D bio-printed human liver tissue stably maintains metabolic functions useful for drug discovery. *Biochem. Biophys. Rep.* **2017**, *10*, 186–191. [CrossRef] [PubMed]

90. Schepers, A.; Li, C.; Chhabra, A.; Seney, B.T.; Bhatia, S. Engineering a perfusable 3D human liver platform from iPS cells. *Lab Chip* **2016**, *16*, 2644–2653. [CrossRef] [PubMed]

91. Shamir, E.R.; Ewald, A.J. Three-dimensional organotypic culture: Experimental models of mammalian biology and disease. *Nat. Rev. Mol. Cell Biol.* **2014**, *15*, 647. [CrossRef] [PubMed]

92. Qian, X.; Nguyen, H.N.; Song, M.M.; Hadiono, C.; Ogden, S.C.; Hammack, C.; Yao, B.; Hamersky, G.R.; Jacob, F.; Zhong, C.; et al. Brain-Region-Specific Organoids Using Mini-bioreactors for Modeling ZIKV Exposure. *Cell* **2016**, *165*, 1238–1254. [CrossRef] [PubMed]

93. Frey, O.; Misun, P.M.; Fluri, D.A.; Hengstler, J.G.; Hierlemann, A. Reconfigurable microfluidic hanging drop network for multi-tissue interaction and analysis. *Nat. Commun.* **2014**, *5*, 4250. [CrossRef] [PubMed]

94. Tung, Y.C.; Hsiao, A.Y.; Allen, S.; Torisawa, Y.S.; Ho, M.; Takayama, S. High-throughput 3D spheroid culture and drug testing using a 384 hanging drop array. *Analyst* **2011**, *136*, 473–478. [CrossRef] [PubMed]

95. Norona, L.M.; Nguyen, D.G.; Gerber, D.A.; Presnell, S.C.; Lecluyse, E.L. Editor's Highlight: Modeling Compound-Induced Fibrogenesis In Vitro Using Three-Dimensional Bioprinted Human Liver Tissues. *Toxicol. Sci.* **2016**, *154*, 354–367. [CrossRef] [PubMed]

96. Kwon, R.Y.; Meays, D.R.; Tang, W.J.; Frangos, J.A. Microfluidic Enhancement of Intramedullary Pressure Increases Interstitial Fluid Flow and Inhibits Bone Loss in Hindlimb Suspended Mice. *J. Bone Miner. Res.* **2010**, *25*, 1798–1807. [CrossRef] [PubMed]

97. Chung, B.G.; Flanagan, L.A.; Rhee, S.W.; Schwartz, P.H.; Lee, A.P.; Monuki, E.S.; Jeon, N.L. Human neural stem cell growth and differentiation in a gradient-generating microfluidic device. *Lab Chip* **2005**, *5*, 401–406. [CrossRef]

98. Andersson, H.; Van Den Berg, A. Microfabrication and microfluidics for tissue engineering: State of the art and future opportunities. *Lab Chip* **2004**, *4*, 98–103. [CrossRef]

99.   Takayama, S.; Ostuni, E.; LeDuc, P.; Naruse, K.; Ingber, D.E.; Whitesides, G.M. Subcellular positioning of small molecules. *Nature* **2001**, *411*, 1016. [CrossRef] [PubMed]

100. Saxena, V.; Nyberg, L.; Pauly, M.; Dasgupta, A.; Nyberg, A.; Piasecki, B.; Winston, B.; Redd, J.; Ready, J.; Terrault, N.A. Safety and Efficacy of Simeprevir/Sofosbuvir in Hepatitis C-Infected Patients with Compensated and Decompensated Cirrhosis. *Hepatology* **2015**, *62*, 715–725. [CrossRef] [PubMed]

101. Geerts, A. History, heterogeneity, developmental biology, and functions of quiescent hepatic stellate cells. *Semin. Liver Dis.* **2001**, *21*, 311–336. [CrossRef] [PubMed]

102. Prot, J.M.; Aninat, C.; Griscom, L.; Razan, F.; Guillouzo, C.G.; Corlu, A.; Brochot, C.; Legallais, C.; Leclerc, E. Improvement of HepG2/C3a cell functions in a microfluidic biochip. *Biotechnol. Bioeng.* **2011**, *108*, 1704–1715. [CrossRef] [PubMed]

103. Kim, C.; Kasuya, J.; Jeon, J.; Chung, S.; Kamm, R.D. A quantitative microfluidic angiogenesis screen for studying anti-angiogenic therapeutic drugs. *Lab Chip* **2015**, *15*, 301–310. [CrossRef] [PubMed]

104. Korin, N.; Kanapathipillai, M.; Matthews, B.D.; Crescente, M.; Brill, A.; Mammoto, T.; Ghosh, K.; Jurek, S.; Bencherif, S.A.; Bhatta, D.; et al. Shear-activated nanotherapeutics for drug targeting to obstructed blood vessels. *Science* **2012**, *337*, 738–742. [CrossRef] [PubMed]

105. Kang, Y.B.A.; Sodunke, T.R.; Lamontagne, J.; Cirillo, J.; Rajiv, C.; Bouchard, M.J.; Noh, M. Liver sinusoid on a chip: Long-term layered co-culture of primary rat hepatocytes and endothelial cells in microfluidic platforms. *Biotechnol. Bioeng.* **2015**, *112*, 2571–2582. [CrossRef] [PubMed]

106. Zhou, Q.; Patel, D.; Kwa, T.; Haque, A.; Matharu, Z.; Stybayeva, G.; Gao, Y.; Diehl, A.M.; Revzin, A. Liver injury-on-a-chip: Microfluidic co-cultures with integrated biosensors for monitoring liver cell signaling during injury. *Lab Chip* **2015**, *15*, 4467–4478. [CrossRef]

107. Kietzmann, T. Metabolic zonation of the liver: The oxygen gradient revisited. *Redox Biol.* **2017**, *11*, 622–630. [CrossRef] [PubMed]

108. Allen, J.W.; Bhatia, S.N. Formation of steady-state oxygen gradients in vitro: Application to liver zonation. *Biotechnol. Bioeng.* **2003**, *82*, 253–262. [CrossRef]

109. McCarty, W.J.; Usta, O.B.; Yarmush, M.L. A microfabricated platform for generating physiologically-relevant hepatocyte zonation. *Sci. Rep.* **2016**, *6*, 26868. [CrossRef] [PubMed]

110. Bhise, N.S.; Manoharan, V.; Massa, S.; Tamayol, A.; Ghaderi, M.; Miscuglio, M.; Lang, Q.; Zhang, Y.S.; Shin, S.R.; Calzone, G.; et al. A liver-on-a-chip platform with bioprinted hepatic spheroids. *Biofabrication* **2016**, *8*, 014101. [CrossRef] [PubMed]

111. Chang, R.; Emami, K.; Wu, H.; Sun, W. Biofabrication of a three-dimensional liver micro-organ as anin vitrodrug metabolism model. *Biofabrication* **2010**, *2*, 045004. [CrossRef] [PubMed]

112. Novik, E.; Maguire, T.J.; Chao, P.; Cheng, K.C.; Yarmush, M.L. A microfluidic hepatic coculture platform for cell-based drug metabolism studies. *Biochem. Pharmacol.* **2010**, *79*, 1036–1044. [CrossRef] [PubMed]

113. Cho, C.H.; Park, J.; Tilles, A.W.; Berthiaume, F.; Toner, M.; Yarmush, M.L. Layered patterning of hepatocytes in co-culture systems using microfabricated stencils. *Biotechniques* **2010**, *48*, 47–52. [CrossRef] [PubMed]

114. Vernetti, L.A.; Senutovitch, N.; Boltz, R.; DeBiasio, R.; Shun, T.Y.; Gough, A.; Taylor, D.L. A human liver microphysiology platform for investigating physiology, drug safety, and disease models. *Exp. Biol. Med. (Maywood)* **2016**, *241*, 101–114. [CrossRef] [PubMed]

115. Wong, S.F.; No, D.Y.; Choi, Y.Y.; Kim, D.S.; Chung, B.G.; Lee, S.H. Concave microwell based size-controllable hepatosphere as a three-dimensional liver tissue model. *Biomaterials* **2011**, *32*, 8087–8096. [CrossRef] [PubMed]

116. Holt, E.; Lunde, P.K.; Sejersted, O.M.; Christensen, G. Electrical stimulation of adult rat cardiomyocytes in culture improves contractile properties and is associated with altered calcium handling. *Basic Res. Cardiol.* **1997**, *92*, 289–298. [CrossRef] [PubMed]

117. Karageorgiou, V.; Kaplan, D. Porosity of 3D biomaterial scaffolds and osteogenesis. *Biomaterials* **2005**, *26*, 5474–5491. [CrossRef] [PubMed]

118. Lee, K.W.; Wang, S.; Dadsetan, M.; Yaszemski, M.J.; Lu, L. Enhanced cell ingrowth and proliferation through three-dimensional nanocomposite scaffolds with controlled pore structures. *Biomacromolecules* **2010**, *11*, 682–689. [CrossRef]

119. Nguyen, A.K.; Narayan, R.J. Two-photon polymerization for biological applications. *Mater. Today* **2017**, *20*, 314–322. [CrossRef]

120. Ovsianikov, A.; Gruene, M.; Pflaum, M.; Koch, L.; Maiorana, F.; Wilhelmi, M.; Haverich, A.; Chichkov, B. Laser printing of cells into 3D scaffolds. *Biofabrication* **2010**, *2*, 014104. [CrossRef] [PubMed]

121. Rekštyte, S.; Kaziulionyte, E.; Balciunas, E.; Kaškelyte, D.; Malinauskas, M. Direct laser fabrication of composite material 3D microstructured scaffolds. *JLMN J. Laser Micro Nanoeng.* **2014**, *9*. [CrossRef]

122. Ovsianikov, A.; Deiwick, A.; Van Vlierberghe, S.; Pflaum, M.; Wilhelmi, M.; Dubruel, P.; Chichkov, B. Laser Fabrication of 3D Gelatin Scaffolds for the Generation of Bioartificial Tissues. *Materials* **2011**, *4*, 288–299. [CrossRef] [PubMed]

123. Zheng, Y.C.; Zhao, Y.Y.; Zheng, M.L.; Chen, S.L.; Liu, J.; Jin, F.; Dong, X.Z.; Zhao, Z.S.; Duan, X.M. Cucurbit uril-Carbazole Two-Photon Photoinitiators for the Fabrication of Biocompatible Three-Dimensional Hydrogel Scaffolds by Laser Direct Writing in Aqueous Solutions. *ACS Appl. Mater. Interfaces* **2019**, *11*, 1782–1789. [CrossRef] [PubMed]

124. Koroleva, A.; Gittard, S.; Schlie, S.; Deiwick, A.; Jockenhoevel, S.; Chichkov, B. Fabrication of fibrin scaffolds with controlled microscale architecture by a two-photon polymerization–micromolding technique. *Biofabrication* **2012**, *4*, 015001. [CrossRef] [PubMed]

125. Du, Y.; Liu, H.; Yang, Q.; Wang, S.; Wang, J.; Ma, J.; Noh, I.; Mikos, A.G.; Zhang, S. Selective laser sintering scaffold with hierarchical architecture and gradient composition for osteochondral repair in rabbits. *Biomaterials* **2017**, *137*, 37–48. [CrossRef] [PubMed]

126. Lohfeld, S.; Tyndyk, M.A.; Cahill, S.; Flaherty, N.; Barron, V.; McHugh, P.E. A method to fabricate small features on scaffolds for tissue engineering via selective laser sintering. *J. Biomed. Sci. Eng.* **2010**, *3*, 138–147. [CrossRef]

127. Negro, A.; Cherbuin, T.; Lutolf, M.P. 3D Inkjet Printing of Complex, Cell-Laden Hydrogel Structures. *Sci. Rep.* **2018**, *8*, 17099. [CrossRef] [PubMed]

128. Mi, S.; Yang, S.; Liu, T.; Du, Z.; Xu, Y.; Li, B.; Sun, W. A Novel Controllable Cell Array Printing Technique on Microfluidic Chips. *IEEE Trans. Biomed. Eng.* **2019**, *66*, 2512–2520. [CrossRef] [PubMed]

129. Derby, B. Bioprinting: Inkjet printing proteins and hybrid cell-containing materials and structures. *J. Mater. Chem.* **2008**, *18*, 5717–5721. [CrossRef]

130. Saunders, R.E.; Derby, B. Inkjet printing biomaterials for tissue engineering: Bioprinting. *Int. Mater. Rev.* **2014**, *59*, 430–448. [CrossRef]

131. Roth, E.A.; Xu, T.; Das, M.; Gregory, C.; Hickman, J.J.; Boland, T. Inkjet printing for high-throughput cell patterning. *Biomaterials* **2004**, *25*, 3707–3715. [CrossRef] [PubMed]

132. Xu, T.; Baicu, C.; Aho, M.; Zile, M.; Boland, T. Fabrication and characterization of bio-engineered cardiac pseudo tissues. *Biofabricatioin* **2009**, *1*, 035001. [CrossRef] [PubMed]

133. Cui, X.; Boland, T. Human microvasculature fabrication using thermal inkjet printing technology. *Biomaterials* **2009**, *30*, 6221–6227. [CrossRef] [PubMed]

134. Zhang, C.; Wen, X.; Vyavahare, N.R.; Boland, T. Synthesis and characterization of biodegradable elastomeric polyurethane scaffolds fabricated by the inkjet tech-nique. *Biomaterials* **2008**, *29*, 3781–3791. [CrossRef] [PubMed]

135. Delaney, J.T.; Liberski, A.R.; Perelaer, J.; Schubert, U.S. Reactive inkjet printing of calcium alginate hydro- gel porogens—A new strategy to open-pore structured matrices with controlled geometry. *Soft Matter* **2010**, *6*, 866–869. [CrossRef]

136. Di Biase, M.; Saunders, R.E.; Tirelli, N.; Derby, B. Inkjet printing and cell seeding thermoreversible photocur-able gel structures. *Soft Matter* **2011**, *7*, 2639–2646. [CrossRef]

137. Xu, T.; Gregory, C.A.; Molnar, P.; Cui, X.; Jalota, S.; Bhaduri, S.B.; Boland, T. Viability and electrophysiology of neural cell structures generated by the inkjet printing method. *Biomaterials* **2006**, *27*, 3580–3588. [CrossRef] [PubMed]

138. Lewis, P.L.; Green, R.M.; Shah, R.N. 3D-printed gelatin scaffolds of differing pore geometry modulate hepatocyte function and gene expression. *Acta Biomater.* **2018**, *69*, 63–70. [CrossRef] [PubMed]

139. Kang, K.; Kim, Y.; Jeon, H.; Lee, S.B.; Kim, J.S.; Park, S.A.; Kim, W.D.; Yang, H.M.; Kim, S.J.; Jeong, J.; et al. Three-Dimensional Bioprinting of Hepatic Structures with Directly Converted Hepatocyte-Like Cells. *Tissue Eng. Part A* **2018**, *24*, 576–583. [CrossRef] [PubMed]

140. Jeon, H.; Kang, K.; Park, S.A.; Kim, W.D.; Paik, S.S.; Lee, S.H.; Jeong, J.; Choi, D. Generation of Multilayered 3D Structures of HepG2 Cells Using a Bio-printing Technique. *Gut Liver* **2017**, *11*, 121–128. [CrossRef] [PubMed]

141. Gong, H.; Agustin, J.; Wootton, D.; Zhou, J.G. Biomimetic design and fabrication of porous chitosan–gelatin liver scaffolds with hierarchical channel network. *J. Mater. Sci. Mater. Med.* **2014**, *25*, 113–120. [CrossRef] [PubMed]

bioengineering

MDPI

Article

3D Printing for Bio-Synthetic Biliary Stents

Christen J. Boyer [1,2], Moheb Boktor [3], Hrishikesh Samant [3], Luke A. White [1], Yuping Wang [4], David H. Ballard [5], Robert C. Huebert [6], Jennifer E. Woerner [2], Ghali E. Ghali [2] and Jonathan S. Alexander [1,*]

[1] Molecular and Cellular Physiology, Health Sciences Center, Louisiana State University, Shreveport, LA 71103, USA; cboye2@lsuhsc.edu (C.J.B.); lwhit9@lsuhsc.edu (L.A.W.)
[2] Oral and Maxillofacial Surgery, Health Sciences Center, Louisiana State University, Shreveport, LA 71103, USA; jwoern@lsuhsc.edu (J.E.W.); GGhali@lsuhsc.edu (G.E.G.)
[3] Gastroenterology and Hepatology, Health Sciences Center, Louisiana State University, Shreveport, LA 71103, USA; mbokto@lsuhsc.edu (M.B.); hsaman@lsuhsc.edu (H.S.)
[4] Obstetrics and Gynecology, Health Sciences Center, Louisiana State University, Shreveport, LA 71103, USA; YWang1@lsuhsc.edu
[5] Mallinckrodt Institute of Radiology, School of Medicine, Washington University, St. Louis, MO 63110, USA; davidballard@wustl.edu
[6] Gastroenterology and Hepatology, Mayo Clinic, Rochester, MN 55905, USA; Huebert.Robert@mayo.edu
* Correspondence: jalexa@lsuhsc.edu; Tel.: +1-318-675-4151; Fax: +1-318-675-6005

Received: 8 January 2019; Accepted: 6 February 2019; Published: 9 February 2019

Abstract: Three-dimensional (3D) printing is an additive manufacturing method that holds great potential in a variety of future patient-specific medical technologies. This project validated a novel crosslinked polyvinyl alcohol (XL-PVA) 3D printed stent infused with collagen, human placental mesenchymal stem cells (PMSCs), and cholangiocytes. The biofabrication method in the present study examined 3D printing and collagen injection molding for rapid prototyping of customized living biliary stents with clinical applications in the setting of malignant and benign bile duct obstructions. XL-PVA stents showed hydrophilic swelling and addition of radiocontrast to the stent matrix improved radiographic opacity. Collagen loaded with PMSCs contracted tightly around hydrophilic stents and dense choloangiocyte coatings were verified through histology and fluorescence microscopy. It is anticipated that design elements used in these stents may enable appropriate stent placement, provide protection of the stent-stem cell matrix against bile constituents, and potentially limit biofilm development. Overall, this approach may allow physicians to create personalized bio-integrating stents for use in biliary procedures and lays a foundation for new patient-specific stent fabrication techniques.

Keywords: 3D printing; hepatobiliary stent; tissue engineering; medical device; stem cells; personalized medicine

---

1. Introduction

Biliary obstruction can be caused by both benign and malignant conditions including iatrogenic bile duct injury and chronic pancreatitis as well as cholangiocarcinoma and pancreatic cancer [1,2]. Relief from such obstructions can help minimize obstructive jaundice and reduce the risk of cholangitis [3,4]. The endoscopic placement of hepatobiliary stents was first performed in 1980 to restore biliary flow [5,6]. Since then, such endoscopic techniques have become favored alternatives to surgery due to their less invasive nature and decreased morbidity associated with these procedures [7].

Several types of stents have been developed over the years, each with their own advantages and disadvantages. Plastic stents were used first and today are still the most commonly used device [5,6]. In the late 1980s, self-expanding metal stents were first used in the biliary tract [8–10]. Compared to plastic stents, self-expanding metal stents have decreased mortality as well as the number of re-interventions, however, self-expanding metal stents are more expensive [4].

More recently, drug-eluting stents have been developed as an approach to maintain stent patency [11]. Both plastic and metal stents with anti-reflux valves have been explored as a means to decrease cholangitis and increase patency [12]. The use of biodegradable biliary stents has also been explored. In many clinical scenarios, biliary stents are only needed temporarily [13,14]. A biodegradable stent that dissolves allowing its remnants to be expelled into the duodenum would eliminate the need for a secondary device removal procedure, which would increase the risk for recurrent biliary occlusion or stenosis.

Despite many advances in stent development, one major problem that remains is the progressive loss of stent patency over time. Several factors contribute to this loss of patency. It is widely accepted that biofilm formation on the stent is a common cause of stent failure. Bacteria gain access to biliary stents by migrating from either the portal venous system or through the sphincter of Oddi [15]. Microorganisms may then adhere to the stent and begin to proliferate and secrete exopolysaccharides to form biofilms that resist antibiotic treatments. Biofilm creation, in turn, is thought to contribute to biliary "sludge" formation that ultimately leads to the loss of stent patency [16].

One area that remains unexplored is the development and use of 3D printed hepatobiliary stents. 3D printing is an emerging technology in medicine, and it would allow for the rapid production of custom medical products that are relatively inexpensive [17]. The application of 3D printed stents has been explored for use in many diseases. Several examples include the use of 3D printed vascular stents for percutaneous coronary intervention, airway stents for the treatment of central airway obstruction, esophageal stents for palliative care in inoperable esophageal malignancies, and ureteral stents which facilitate urine drainage from the kidney [18–21].

Cytocompatible, hydrophilic, and drug-deliverable 3D printed materials and techniques have been recently explored and hold great promise in cell culture models, medical devices, and tissue-engineered substrates [22–25]. Such materials include 3D printed polyvinyl alcohol (PVA), a water soluble polymer. PVA filaments demonstrate favorable 3D printability using fused deposition modeling (FDM) techniques which can be post-print crosslinked (XL) to form stable 3D hydrogels [26]. The 3D printed crosslinked polyvinyl alcohol (XL-PVA) method is capable of binding biomolecules, drugs, and cells of interest due to the molecular complexes formed during crosslinking. Such materials have applications in a variety of medical devices and wound care substrates. In this study, the XL-PVA 3D printing method was coupled with collagen injection molding to create hepatobiliary stents which have been engineered to support engraftment with human placental stem cells and cholangiocytes.

2. Materials and Methods

2.1. PVA Stent Fabrication and Crosslinking

Porous tubular stents (25 mm length, 5 mm external diameter, and 3 mm internal diameter) were created with free computer aided design (CAD) software (Tinker-CAD, AutoDesk, San Francisco, CA, USA). 96 square pores (1 mm × 1 mm) were created in the stent design with 1 mm spacing between each pore. PVA filaments (AquaSolve™, Formfutura, Nijmegen, Netherlands, 1.75 mm diameter) were 3D printed into the designed stent pattern at 201 °C using a consumer grade 3D printed (MakerBot Replicator desktop 3D printer, MakerBot Industries LLC, Brooklyn, NY, USA) with supports turned off and raft turned on. PVA 3D printed stents were immersed in distilled water briefly to fuse layers, and cross-linked (XL) by placing the stents in a gas vapor desiccator with two separate containers containing 20 mL of 6.25% glutaraldehyde (GA) (EMD Millipore Corporation, Darmstadt, Germany) and 10 mL of concentrated hydrochloric acid (HCl) (Fisher Scientific, Hampton, NH, USA) at 42 °C for 24 h. XL-PVA stents were next rinsed extensively in distilled water and soaked in 70% ethanol for 24 h. XL-PVA stents were rinsed again and placed in 1X phosphate buffer solution (PBS) for storage.

2.2. Collagen Injection Mold Chamber Fabrication

Cylindrical injection mold chambers and stent lumen maturation plugs were designed with free CAD software (Tinker-CAD) and 3D printed on a consumer grade stereolithography 3D printer (Formlabs, Form2, Somerville, MA, USA) using flexible resin (Formlabs) with supports turned on. 3D printed parts were rinsed in 70% isopropyl alcohol (Form Wash, Formlabs) and then cured at 60 °C with shortwave ultraviolet light (UV) for 1 h (Form Cure, Formlabs). Injection mold chambers and maturation plugs were next rinsed in 70% ethanol followed by a $1\times$ PBS wash in a sterile tissue culture hood and stored there until used in experiments.

2.3. Barium Sulfate Coating and X-Ray Imaging

Polycaprolactone (PCL) (Sigma Aldrich, St. Louis, MO, USA) (molecular weight 80,000) and E-Z-HD™ barium sulfate (BS) (E-Z-EM Canada Inc., Quebec, Canada) were mixed in chloroform (CF) (Sigma Aldrich) to form 10% PCL:10% BS:80% CF mixture. Hydrated XL-PVA stents were dipped in the PCL-BS-CF mixtures on each tip's end and the chloroform was allowed to evaporate before placing the stents back into $1\times$ PBS. Final dry PCL-BS coatings on stent tips contained 50% PCL:50% BS. XL-PVA stents and versions with PCL-BS coatings were imaged with an OEC 9900 Elite C-Arm System X-ray (General Electric, Fairfield, CT, USA) in 10 mL of $1\times$ PBS.

2.4. XL-PVA Swelling Analysis

For swelling studies, XL-PVA pieces (n = 5) were hydrated overnight in 1X PBS and dehydrated. Each hydrated and dehydrated form was weighed to obtain average weights. To calculate mass change percentage (ΔM%), the average hydrated weight (M1) was subtracted from the average dry weight (M2), then divided by the hydrated weight (M1) and multiplied by 100. The mass swelling ratio was calculated by taking the ratio between the average mass of the hydrated XL-PVA (M1) and the average mass of the dehydrated XL-PVA (M2), where M1 is divided by M2. The results were statistically analyzed by using Student's t-test and the data were expressed as the mean $\pm$ standard deviation (SD). A p-value of <0.05 was considered statistically significant.

2.5. Collagen Preparation

Rat tail type 1 collagen matrices were prepared by a modification of the protocol previously published by Benoit et al. [27]. Briefly, rat tail tendons were manually excised, washed with 100% isopropanol (Thermo Fisher Scientific, Waltham, MA, USA) and dissolved in sterile 4 mM acetic acid (Sigma Aldrich) for 24 h at 4 °C under constant agitation. Collagen solution was filtered through a 250 μm nylon filter (Spectrum Labs, Rancho Dominguez, CA, USA), centrifuged at $19\times$ g for 20 min at 4 °C and snap frozen. Using a bench-top manifold freeze-dryer (Millrock Technology, Kingston, NY, USA), frozen aliquots were dried and stored at -20 °C for future use. Twenty-four hours prior to experiments, freeze-dried collagen was resolubilized in cold 0.012 M hydrochloric acid (HCl) (Sigma Aldrich) at 2.5 mg/mL final collagen concentration and incubated overnight at 4 °C with gentle agitation. On the day of the experiment, 0.8 mL of cold 5X PBS was added to 3.2 mL of dissolved collagen gel and the pH was titrated with 0.5 M sodium hydroxide (NaOH) (Sigma Aldrich) to 7.4.

2.6. Isolation of Mesenchymal Stem Cells

Human placentas were collected from normal term pregnancies. Collection of human placenta for mesenchymal stem cell isolation was approved by the Institutional Review Board (IRB) (approval number CR00001345_STUDY00000614). Human placental mesenchymal stem cells (PMSCs) were obtained by cultivation of microvilli after elimination of villous trophoblasts. The villous tissue was dissected from different cotyledons, excluding chorionic and basal plates. After rinsing with ice-cold PBS, the villous tissue was next digested with trypsin (0.125% trypsin solution (Sigma Aldrich) containing 0.1 mg/mL DNase I and 5 (Sigma Aldrich) mM $MgCl_2$ (Sigma Aldrich) in Dulbecco's modified Eagle's medium (DMEM) (Sigma Aldrich) at 37 °C for 90 min. The isolated PMSCs were incubated with DMEM supplemented with 10% fetal bovine serum (FBS) (Atlanta Biological, Flowery Branch, GA, USA) and 1% penicillin/streptomycin (P/S) (Sigma Aldrich) at 37 °C with 5% CO_2. PMSCs were characterized by positive expression of mesenchymal stem cell cluster of differentiation (CD) markers CD73, CD90, and negative expression of CD34 and Human Leukocyte Antigen–antigen D Related (HLA-DR) (BD Pharmingen, Franklin Lakes, NJ, USA). PMSCs were monitored using flow cytometry (BD LSR II Flow Cytometer, BD Biosciences, San Jose, CA, USA) and immunofluorescent staining for octamer binding transcription factor 4 (Oct4), CD133 (Santa Cruz, San Diego, CA, USA) and CD44 (Abcam, Cambridge, MA, USA).

2.7. Stem Cell Collagen Injection Molding and Stent Maturation

All cell culture was performed under standard aseptic techniques to reduce contamination. PMSC (passage 9) was grown to confluency in 75 cm^2 flat bottom flasks at 37 °C with 7.5% carbon dioxide (CO_2), and 100% humidity. Cells were washed with PBS/ethylenediaminetetraacetic acid (PBS-EDTA) (Sigma Aldrich), trypsinized (Trypsin solution from porcine pancreas, Sigma Aldrich), collected through centrifugation, and re-suspended in DMEM (media) containing 10% FBS, (Atlas Biologicals, Fort Collins, CO, USA), 1% P/S (Sigma Aldrich), and 4.5 g of glucose/liter. Resuspended PMSCs were mixed with the prepared dissolved collagen solution to form a 12 mL mixture (8 mL of cell suspension in 4 mL of collagen gel solution). PMSC/collagen was seeded in the 3D printed injection molding chambers containing XL-PVA stents and maturation plugs. 2 mL of the PMSC-collagen mixture was added to each injection molding chamber and the total cell count for each chamber was 200,000 cells per chamber. The loaded injection molding chambers tissue were next incubated at 37 °C for 1 h., to polymerize the collagen around the 3D prints. Next, maturation plugs containing XL-PVA stents and polymerized collagen/PMSC were removed from each injection molding chamber and placed in 25 cm^2 tissue culture flask with 25 mL of media. Maturation plugs were removed at day 5 and PMSC/collagen stents were allowed to mature for a total of 7 days.

Additional 48-well tissue culture plates were seeded with the PMSC-collagen mixtures and pure collagen as a control. Each well was seeded with 0.5 mL of the solutions and PMSC versions contained 50,000 cells per well. The loaded 48 well tissue culture plates were next incubated at 37 °C for 1 h., to polymerize the collagen in the wells. Contraction assays measuring the diameter of the gel were performed according to established protocols [27]. Collagen gels in the 48-well plates were detached around the edges with a glass pipet tip and 0.5 mL of media was added to each well. Collagen contraction was monitored for 1 week, and images were taken periodically on days 0, 3, 5, and 7), and analyzed with ImageJ software. The results were statistically analyzed by using Student's *t*-test, and the data were expressed as the mean $\pm$ standard deviation (SD). A *p*-value of <0.05 was considered statistically significant.

2.8. Cholangiocyte Seeding

Human primary cholangiocytes (Celprogen, Torrance, CA, USA) were grown to confluency in 75 cm^2 tissue culture flask with human cholangiocyte primary cell culture complete growth media with serum and antibiotics (Celprogen). The obtained primary cholangiocytes were stated to have positive markers for cytokeratin (CK) CK7, CK19, glutamyl transpeptidase, aquaporin -1 (Aqp1), oval cell markers (OV) OV6 and OV1, and epithelial specific antigen (ESA). Cells were washed with PBS-EDTA, trypsinized, collected through centrifugation, and re-suspended in Celprogen media. Collagen injection molded XL-PVA stents were each mixed with 5 mL of cholangiocyte media containing 663,000 cell count and allowed to incubate overnight. Stents were next transferred to new 25 cm^2 flask and 25 mL of fresh Celprogen media was added. Stents were allowed to mature for an additional week. Final cell-laden stents were analyzed live and fixed in 10% formalin (Thermo Fisher) for imaging and histology processing. For histology, fixed stents were embedded in parafilm wax, sectioned, and stained with hematoxylin and eosin (H&E) (Sigma Aldrich) for viewing.

2.9. Imaging

3D printed XL-PVA and collagen injection molded versions seeded with cells were characterized with a HITACHI 4800 high resolution scanning electron microscope with Gatan Cryo features (Cryo-SEM). Cell cultures and external and internal cell layers of bioengineered stents were also monitored through phase and fluorescence imaging with Hoechst 33342 (Thermo Fisher) on an Evos FL cell imaging system (Thermo Fisher). H&E stained sections were viewed with an EVOS XL Core imaging system (Thermo Fisher).

3. Results

3.1. Fabrication of XL-PVA Stents and Collagen Injection Molding Chamber

Consumer grade free software and a MakerBot 3D printer were capable of forming reproducible stent structures from PVA filaments printed at 201 °C. (see Figures 1 and 2). 3D printed PVA stents were successfully cross-linked with the HCl/GA gas crosslinking method previously described (see Figure 2D,E). The Form2 SLA 3D printer was successful at fabricating collagen injection molding chambers and maturation plugs (see Figure 2). The injection molding chambers and maturation plugs allowed for smooth integration with each part including the XL-PVA stents (see Figure 2D,E). XL-PVA stents coated with PCL/BA showed improved visibility under x-ray imaging when compared to control XL-PVA stents in PBS (see Figure 3).

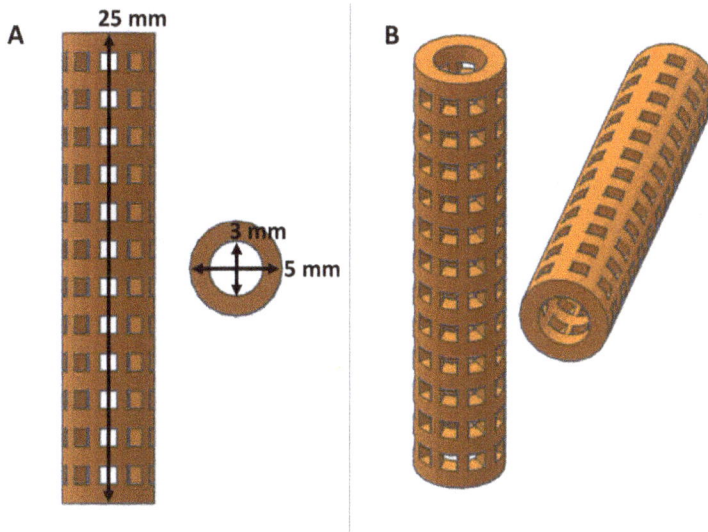

Figure 1. Views of computer aided design (CAD) stent with dimensions displayed in orthographic view (**A**) and in perspective view (**B**).

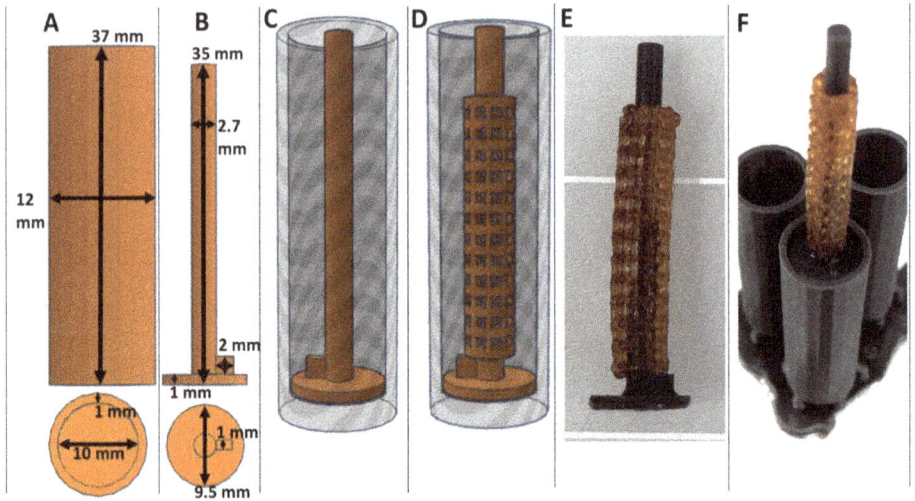

Figure 2. Views of CAD collagen injection molding chamber (**A**) and maturation plug (**B**) with dimensions displayed. CAD collagen injection molding chamber with maturation plug inserted (**C**) and with stent placement (**D**). Images of 3D printed actual crosslinked polyvinyl alcohol (XL-PVA) stents, over a maturation plug (**E**) and with the injection molding chambers (**F**).

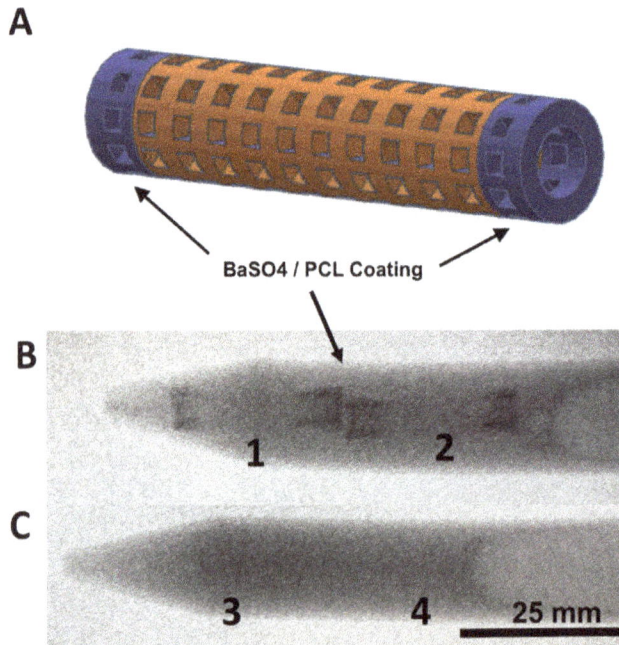

Figure 3. Images of CAD stent in (**A**), X-ray images of 3D printed XL-PVA stents with barium coated tips (**B**; 1 and 2), and X-ray image of control XL-PVA stents (**C**; 3 and 4). Black arrows point to regions of interest containing barium sulfate (scale bar = 25 mm).

3.2. XL-PVA Swelling Analysis

Hydrated samples of XL-PVA (n = 5) had a mean mass (grams) ± standard deviation of 0.4444 ± 0.0220. Dehydrated samples of same XL-PVA samples (n = 5) had a mean ± standard deviation mass (grams) of 0.2734 ± 0.0137. For hydrated versus dehydrated XL-PVA, a significant difference was found for the mass totals ($p < 0.0001$). The change in mass percentage from hydrated to dehydrated XL-PVA was calculated as 38.47% (see Figure 4). The mass swelling ratio for hydrated and dehydrated XL-PVA was calculated as 1.625.

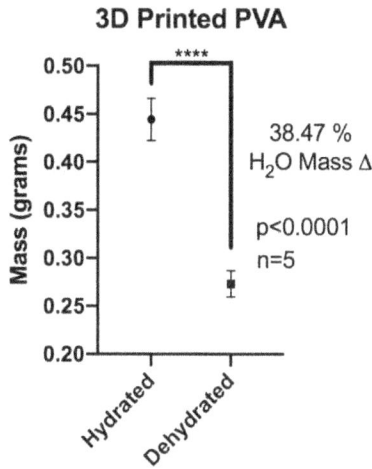

Figure 4. Chart displaying average masses of XL-PVA in hydrated and dehydrated forms. (**** = $p < 0.0001$).

3.3. Hybrid Stent Cell Culture and Imaging

Human PMSCs and cholangiocytes were successfully cultured on stents; these displayed unique morphologies when compared under phase microscopy. PMSCs displayed typical elongated and uniform web-like patterns while the cholangiocytes displayed typical circular and cobblestone patterns (see Figure 5). Under flow cytometry, PMSCs showed positive (+) expression for several markers (CD73+, CD90+) and negative (−) for markers CD34− and HLADR−. Under immunofluorescent imaging PMSCs showed CD133+, CD44+, and Oct4+.

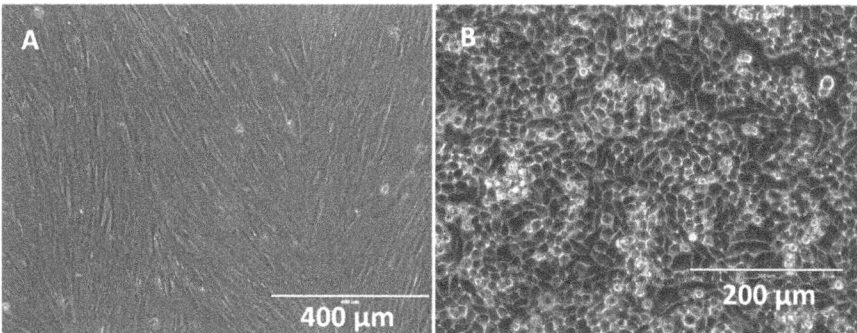

Figure 5. Phase microscopy images of confluent human placental mesenchymal stem cells (PMSCs) (**A**) (scale bar = 400 μm) and cholangiocytes (**B**) (scale bar = 200 μm).

Collagen gels loaded with PMSCs showed significant contraction in 48-well cell culture plates and around the 3D printed XL-PVA stents. In 48-well plates, the diameters of the gels were measured and images were taken at the start of the experiment day 0 and at days 3, 5, and 7. Control collagen gels (n = 9) had a mean ± standard deviation diameter (mm) of 15.943 ± 0.308 for days 0–7 with no contraction observed. PMSC collagen gels (n = 9) (p-values = Student t-test comparison of control collagen gels) versus PMSC gels for each day had a mean ± standard deviation diameter (mm) of 15.800 ± 0.201 for day 0 (p = 0.2632), 11.230 ± 0.590 ($p < 0.0001$) for day 3, 9.230 ± 0.448 ($p < 0.0001$) for day 5, and 7.460 ± 0.630 ($p < 0.0001$) for day 7. No significant difference was observed for day 0 with PMSCs seeded in gels, and each day after (days 3–7) displayed significant differences in average diameter when compared to control gels (see Figure 6). Additionally, each PMSC collagen gel was significantly different when compared to each other on 0–7 days ($p < 0.0001$).

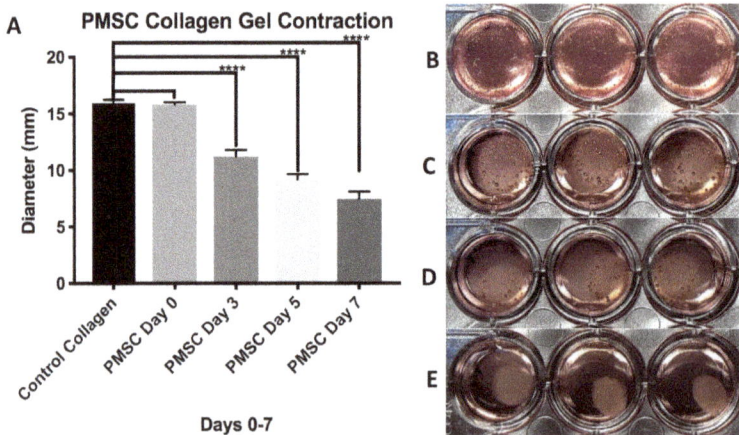

Figure 6. Chart displaying average diameter of control collagen gels and PMSC seeded collagen gels at days 0–7 days (**A**). Images on right show collagen gels seed with PMSCs at day 0 (**B**), day 3 (**C**), day 5 (**D**), and day 7 (**E**). (**** = $p < 0.0001$).

Similarly, PMSC collagen contraction was observed around the XL-PVA stents over the course of 7 days (see Figure 7). Upon removal of the maturation plug from the PMSC collagen seeded stents, patency of the inner lumen was observed and maintained at day 5 and contraction continued to day 7 (see Figure 7C). CryoSEM images showed that both XL-PVA and collagen surfaces promoted cell attachment with cholangiocytes (see Figure 8).

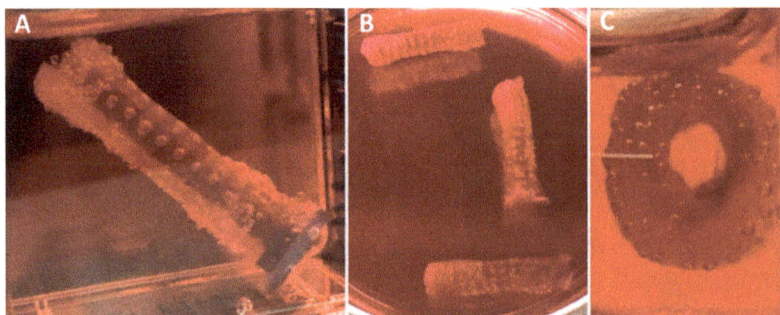

Figure 7. Images of 3D printed XL-PVA stents with maturation plug and collagen/PMSC at day 5 (**A**) and with plug removed (**B**,**C**).

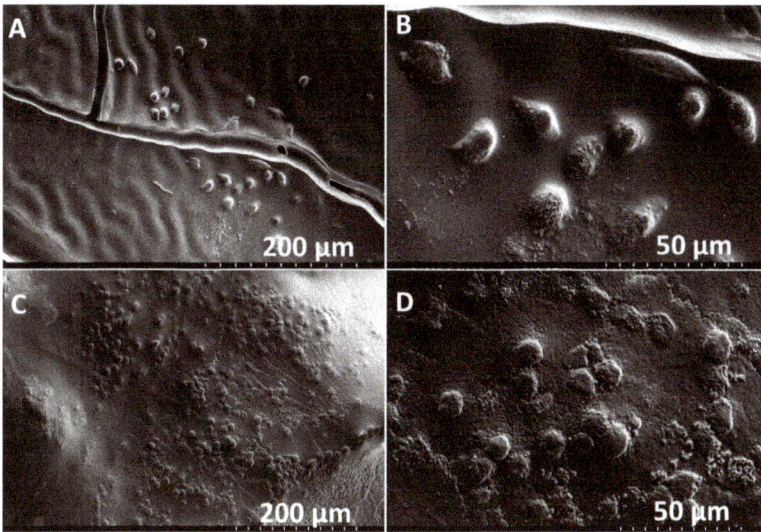

Figure 8. Cryo-SEM images of XL-PVA with cholangiocytes on the surface (**A,B**). Cryo-SEM images of collagen with cholangiocytes on the surface (**C,D**). (**A** and **C**, scale bar = 200 μm) and (**B** and **D**, scale bar = 50 μm).

Stents containing XL-PVA, collagen, PMSC, and cholangiocytes that have matured for two weeks displayed tight uniform collagen coatings around the XL-PVA and densely bound monolayers of cholangiocytes on the surfaces (see Figure 9). Both inner and outer stent surface showed dense cholangiocyte surfaces. Additionally, cross-sections of inner and outer stent surfaces displayed contracted collagen with cholangiocyte monolayers visible by both fluorescent imaging and histology (see Figure 10).

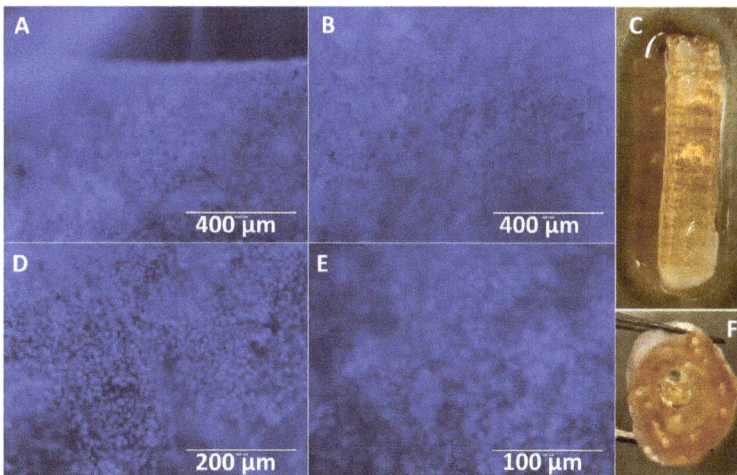

Figure 9. Images of Hoechst 33342 fluorescence (blue) staining of cholangiocytes on the outer stent surface (**A,B**), and image of outer stent (**C**). Images of Hoechst 33342 fluorescence (blue) staining of cholangiocytes on the inner stent surface (**D,E**) and image of inner lumen of stent (**F**) at day 7. (**A** and **B**, scale bar = 400 μm). (**D**, scale bar =200 μm). (**E**, scale bar = 100 μm).

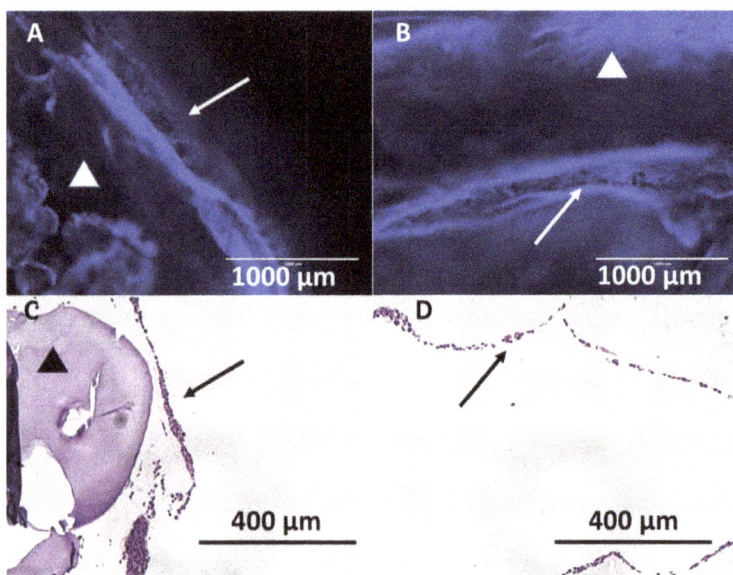

Figure 10. Images of Hoechst 33342 fluorescence (blue) staining of cholangiocytes on the outer stent surface (**A**) and of inner stent surface (**B**). (**A** and **B**, scale bar = 1000 μm). Images of H&E stain of outer stent surface (**C**) and of inner stent surface (**D**). (**C** and **D**, scale bar = 400 μm) (Arrows = cholangiocyte layer, and triangles = 3D printed XL-PVA area).

4. Discussion

In this current proof-of-concept study, both a novel 3D printed biliary stent fabrication technique as well as a method for injection molding and coating these stents with a stem cell and collagen-cholangiocyte linings were reported. In these in vitro studies, predictable swelling of the hydrated stent matrix was achieved, with matrices absorbing nearly an equivalent water mass over the dehydrated cross-linked print. The mass of the stent can be manipulated through hydration, which may be important when deploying the dehydrated stent endoscopically into the wet environment of the common bile duct in vivo. These stents were successfully modified with barium contrast to create x-ray attenuation, an important factor for fluoroscopic visualization and localization during endoscopic procedures. A prior study of impregnated contrast materials showed barium to be more resilient than iodinated contrast materials when incorporated into the structure of 3D-printed constructs (i.e., surgical mesh used in that study) [22]. Additional PCL and other polymer coatings or drugs could be applied to PVA-based stents for a variety of material modifications including increased mechanical properties, enhanced cell attachment, and optimal drug delivery. Additionally, other polymer 3D print materials could be integrated with 3D printed injection molding systems.

The XL-PVA crosslinking method creates a stable and stiff 3D hydrogel with favorable collagen and cell adhesive properties. While the degree of crosslinking was not quantitively measured, it was qualitatively observed that shorter time periods of crosslinking and more solid prints resulted in more unstable stents that fractured randomly upon hydration due to uncross-linked regions. The overall size of the PVA print and macro-porosity in this study allowed for consistent and stable stent structures with good handling upon hydration. The 1 mm porosity allowed for consistent global crosslinking with suitable features for this method. 1 mm dimensions are also compatible with most consumer grade 3D printers. The PMSC-collagen gels in a 48-well configuration and in 3D printed stents showed progressive tonic contraction which contracts around and fuses with the print material. The PMSCs exhibit different adherence morphology in 2D and 3D cultures. PMSCs grown in collagen gels are able to form 3D arrangement and attachment within the collagen network. Cytoskeleton branching is able to form in all directions, which translates to global contraction of the collagen gels. This stem cell contraction is an important feature to consider when designing such collagen frameworks for tissue engineering, and such material combinations are not limited to hepatobiliary stents. A variety of vascular tissues and grafts may also be adapted using this technique with PMSCs as they are readily available non-invasive and non-controversial.

Cholangiocytes displayed similar adherence morphology in 2D and 3D cultures with a tightly bound single layer cobblestone pattern. The uniform cobblestone pattern occurred on both the tissue culture plastics and around the collagen stent as a coating. Both direct microscopic visualization along with fluorescent and histologic staining confirmed the presence of cholangiocytes on the surface of these 3D printed stents. It is hypothesized that the addition of an external cholangiocyte layer is intended to enhance integration and eventually fuse to the inner stent cholangiocyte layer. Biofilms generally occur on the surface of plastics which allow adherence and biofilm formation [15,16]. It is suggested that by using patient cells or other tightly bound cell types, a stent surface could be engineered with cells to reduce entrance and adherence of harmful bacteria. With such new collagen fabrication methods entering the market, such as recombinant human collagen, a very wide variety of biocompatible collagen 3D print-based technologies may be developed in the near future with patient cells [28,29].

Cumulatively, the parameters in this study are relevant for creating, testing and manipulating many 3D printed formats besides biliary stents. The design of this stent is customizable and can be custom-designed to match patient-specific anatomies from medical imaging. Coupled with cryogenic storage and on-demand subtractive sculpting, stents with variable sizes could be achieved through use of these methods [30–33]. Although the incorporation of cholangiocytes onto the 3D printed stent's surface was successfully accomplished, additional work is anticipated, including long-term cell viability and optimal thickness, animal studies, and mechanical testing which will demonstrate how effectively these coatings and materials integrate and maintain stent patency.

5. Conclusions

In conclusion, this study demonstrated in vitro a proof-of-concept synthesis of 3D printed plastic biliary stents impregnated with barium along with a stem cell-collagen-cholangiocyte coating. The biodegradation of the PVA is an ongoing study where we are examining degradation in bile and animal models. We hope to publish future findings on the effects of crosslinking on the degradation of PVA in vitro and in vivo. It is hypothesized that less crosslinked materials will result in faster biodegradation and mechanical degradation. However, they may not be suitable for implantation and handling. Overall, a wide variety of compatible 3D printing materials could be used with this method and it is not limited to crosslinked PVA. To our knowledge, this is the first report of 3D printing technology being used to fabricate custom biliary bio-stents. Whereas the 3D printing allows for customization of the stent structure and impregnation of barium for visibility, the properties of the matrix allows for variations of the mass and swelling of the stent wall composition. These are further modified by the living coating, which promotes contraction and integration of the components of the stent. Aside from the cholangiocytes incorporated on the surface in the present study, the collagen coating process may facilitate other enhancements on the 3D printed biliary stent's surface. All of these factors are in keeping with the novel capabilities of 3D printing to facilitate customizable, patient-specific medicine. These advancements may help facilitate custom biliary stent design, aimed at improving patency and patient care.

6. Patents

Alexander, J.S.; Boyer, C.J. 3D Printed Polyvinyl Alcohol Medical Devices and Methods of Activation. Assignee: LSU Health Sciences Center Shreveport, LA. U.S. Non-Provisional Patent Application 15/721,561. Wang, Y. Digested placental microvilli culture: a simple, efficient, and reproducible process to obtain stromal/mesenchymal stem (stem cell-like) cells from human term

Bioengineering **2019**, *6*, 16

placenta. Assignee: LSU Health Sciences Center Shreveport, LA. U.S. Provisional Patent Application 62/658,084.

Author Contributions: All authors contributed to the development of this research article. Conceptualization, C.J.B., M.B., H.S., and J.S.A.; methodology, C.J.B., M.B., H.S., J.S.A., R.C.H., Y.W.; validation, C.J.B., M.B., H.S., J.S.A.; formal analysis, C.J.B., M.B., H.S., J.S.A.; investigation, C.J.B., M.B., H.S., and J.S.A.; resources, C.J.B., M.B., H.S., J.S.A., Y.W., R.C.H.; data curation, C.J.B., M.B., H.S., J.S.A.; writing—original draft preparation, C.J.B., M.B., H.S., J.A, L.A.W., Y.W., D.H.B., R.C.H., J.E.W., and G.E.G.; writing—review and editing, C.J.B., M.B., H.S., J.A, L.A.W., Y.W., D.H.B., R.C.H., J.E.W., and G.E.G.; visualization, C.J.B., and J.S.A.; supervision, C.J.B., M.B., and J.S.A.; project administration, J.S.A.; funding acquisition, C.J.B., M.B., H.S., and J.S.A.

Funding: This work was supported by the LSU Research and Technology Foundation. (LSU LIFT2 Grant 110351233A). DHB receives salary support from National Institutes of Health TOP-TIER grant T32-EB021955.

Acknowledgments: Authors would like to thank Jibao He, at Tulane University, for assisting with electron microscope imaging.

Conflicts of Interest: Authors declare no conflicts of interest.

References

1. Moy, B.T.; Birk, J.W. An Update to Hepatobiliary Stents. *J. Clin. Transl. Hepatol.* **2015**, *3*, 67–77. [CrossRef] [PubMed]
2. Costamagna, G.; Boškoski, I. Current treatment of benign biliary strictures. *Ann. Gastroenterol.* **2013**, *26*, 37–40. [PubMed]
3. Lee, J.; DaVee, T. Biliary Obstruction: Endoscopic Approaches. *Semin. Interv. Radiol.* **2017**, *34*, 369–375.
4. Dumonceau, J.M.; Tringali, A.; Papanikolaou, I.; Blero, D.; Mangiavillano, B.; Schmidt, A.; Vanbiervliet, G.; Costamagna, G.; Deviere, J.; Garcia-Cano, J. Endoscopic biliary stenting: Indications, choice of stents, and results: European Society of Gastrointestinal Endoscopy (ESGE) Clinical Guideline—Updated October 2017. *Endoscopy* **2018**, *50*, 910–930. [CrossRef] [PubMed]
5. Soehendra, N.; Reynders-Frederix, V. Palliative bile duct drainage—A new endoscopic method of introducing a transpapillary drain. *Endoscopy* **1980**, *12*, 8–11. [CrossRef] [PubMed]
6. Laurence, B.H.; Cotton, P.B. Decompression of malignant biliary obstruction by duodenoscopic intubation of bile duct. *Br. Med. J.* **1980**, *280*, 522–523. [CrossRef] [PubMed]
7. Davids, P.H.; Tanka, A.K.; Rauws, E.A.; Gulik, T.M.; Leeuwen, D.J.; Wit, L.T.; Verbeek, P.C.; Huibregtse, K.; Heyde, M.N.; Tytgat, G.N. Benign biliary strictures. Surgery or endoscopy? *Ann. Surg.* **1993**, *217*, 237–243. [CrossRef] [PubMed]
8. Irving, J.D.; Adam, A.; Dick, R.; Dondelinger, R.F.; Lunderquist, A.; Roche, A. Gianturco expandable metallic biliary stents: Results of a European clinical trial. *Radiology* **1989**, *172*, 321–326. [CrossRef] [PubMed]
9. Neuhaus, H.; Hagenmüller, F.; Classen, M. Self-expanding biliary stents: Preliminary clinical experience. *Endoscopy* **1989**, *21*, 225–228. [CrossRef] [PubMed]
10. Huibregtse, K.; Cheng, J.; Coene, P.P.; Fockens, P.; Tytgat, G.N. Endoscopic placement of expandable metal stents for biliary strictures—A preliminary report on experience with 33 patients. *Endoscopy* **1989**, *21*, 280–282. [CrossRef] [PubMed]
11. Suk, K.T.; Kim, J.W.; Kim, H.S.; Baik, S.K.; Oh, S.J.; Lee, S.J.; Kim, H.G.; Lee, D.H.; Won, Y.H.; Lee, D.K. Human application of a metallic stent covered with a paclitaxel-incorporated membrane for malignant biliary obstruction: Multicenter pilot study. *Gastrointest. Endosc.* **2007**, *66*, 798–803. [CrossRef] [PubMed]
12. Hair, C.D.; Sejpal, D.V. Future developments in biliary stenting. *Clin. Exp. Gastroenterol.* **2013**, *6*, 91–99. [PubMed]
13. Siiki, A.; Rinta-Kiikka, I.; Sand, J.; Laukkarinen, J. Endoscopic biodegradable biliary stents in the treatment of benign biliary strictures: First report of clinical use in patients: Biodegradable stents in ERCP. *Dig. Endosc.* **2017**, *29*, 118–121. [CrossRef] [PubMed]
14. Mauri, G.; Michelozzi, C.; Melchiorre, F.; Poretti, D.; Pedicini, V.; Salvetti, M.; Criado, E.; Falco Fages, J.; De Gregorio, M.A.; Laborda, A. Benign biliary strictures refractory to standard bilioplasty treated using polydoxanone biodegradable biliary stents: Retrospective multicentric data analysis on 107 patients. *Eur. Radiol.* **2016**, *26*, 4057–4063. [CrossRef] [PubMed]

15. Sung, J.Y.; Leung, J.W.C.; Shaffer, E.A.; Lam, K.; Olson, M.E.; Costerton, J.W. Ascending infection of the biliary tract after surgical sphincterotomy and biliary stenting. *J. Gastroenterol. Hepatol.* **1992**, *7*, 240–245. [CrossRef] [PubMed]

16. Guaglianone, E.; Cardines, R.; Vuotto, C.; Di Rosa, R.; Babini, V.; Mastrantonio, P.; Donelli, G. Microbial biofilms associated with biliary stent clogging. *FEMS Immunol. Med. Microbiol.* **2010**, *59*, 410–420. [CrossRef] [PubMed]

17. Ventola, C.L. Medical Applications for 3D Printing: Current and Projected Uses. *P T Peer-Rev. J. Formul. Manag.* **2014**, *39*, 704–711.

18. Misra, S.K.; Ostadhossein, F.; Babu, R.; Kus, J.; Tankasala, D.; Sutrisno, A.; Walsh, K.A.; Bromfield, C.R.; Pan, D. 3D-Printed Multidrug-Eluting Stent from Graphene-Nanoplatelet-Doped Biodegradable Polymer Composite. *Adv. Healthc. Mater.* **2017**, *6*, 11. [CrossRef] [PubMed]

19. Xu, J.; Ong, H.X.; Traini, D.; Byrom, M.; Williamson, J.; Young, P.M. The utility of 3D-printed airway stents to improve treatment strategies for central airway obstructions. *Drug Dev. Ind. Pharm.* **2018**, *45*, 1–10. [CrossRef] [PubMed]

20. Lin, M.; Firoozi, N.; Tsai, C.T.; Wallace, M.B.; Kang, Y. 3D-printed flexible polymer stents for potential applications in inoperable esophageal malignancies. *Acta Biomater.* **2018**, *83*, 119–129. [CrossRef] [PubMed]

21. Del Junco, M.; Yoon, R.; Okhunov, Z.; Abedi, G.; Hwang, C.; Dolan, B.; Landman, J. Comparison of Flow Characteristics of Novel Three-Dimensional Printed Ureteral Stents Versus Standard Ureteral Stents in a Porcine Model. *J. Endourol.* **2015**, *29*, 1065–1069. [CrossRef] [PubMed]

22. Ballard, D.H.; Jammalamadaka, U.; Tappa, K.; Weisman, J.A.; Boyer, C.J.; Alexander, J.S.; Woodard, P.K. 3D printing of surgical hernia meshes impregnated with contrast agents: In vitro proof of concept with imaging characteristics on computed tomography. *3D Print. Med.* **2018**, *4*, 13. [CrossRef] [PubMed]

23. Boyer, C.J.; Ballard, D.H.; Barzegar, M.; Yun, J.W.; Woerner, J.E.; Ghali, G.E.; Boktor, M.; Wang, Y.; Alexander, J.S. High-throughput scaffold-free microtissues through 3D printing. *3D Print. Med.* **2018**, *4*, 9. [CrossRef] [PubMed]

24. Boyer, C.J.; Ballard, D.H.; Yun, J.W.; Xiao, A.Y.; Weisman, J.A.; Barzegar, M.; Alexander, J.S. Three-dimensional printing of cell exclusion spacers (CES) for use in motility assays. *Pharm. Res.* **2018**, *35*, 155. [CrossRef] [PubMed]

25. Boyer, C.J.; Galea, C.; Woerner, J.E.; Gatlin, C.A.; Ghali, G.E.; Mills, D.K.; Weisman, J.A.; McGee, D.J.; Alexander, J.S. Personalized bioactive nasal supports for postoperative cleft rhinoplasty. *J. Oral Maxillofac. Surg.* **2018**, *76*, 1562.e1–1562.e5. [CrossRef] [PubMed]

26. Boyer, C.J.; Ballard, D.H.; Weisman, J.A.; Hurst, S.; McGee, D.J.; Mills, D.K.; Woerner, J.E.; Jammalamadaka, U.; Tappa, K.; Alexander, J.S. Three-Dimensional Printing Antimicrobial and Radiopaque Constructs. *3D Print Addit. Manuf.* **2018**, *5*, 29–35. [CrossRef]

27. Benoit, C.; Gu, Y.; Zhang, Y.; Alexander, J.S.; Wang, Y. Contractility of placental vascular smooth muscle cells in response to stimuli produced by the placenta: Roles of ACE vs. non-ACE and AT1 vs. AT2 in placental vessel cells. *Placenta* **2008**, *29*, 503–509. [CrossRef] [PubMed]

28. Shoseyov, O.; Posen, Y.; Grynspan, F. Human Recombinant Type I Collagen Produced in Plants. *Tissue Eng. Part A* **2013**, *19*, 1527–1533. [CrossRef] [PubMed]

29. Yaari, A.; Schilt, Y.; Tamburu, C.; Raviv, U.; Shoseyov, O. Wet Spinning and Drawing of Human Recombinant Collagen. *ACS Biomat. Sci. Eng.* **2016**, *2*, 349–360. [CrossRef]

30. Trunec, M.; Chlup, Z. Substractive manufacturing of customized hydroxyapatite scaffolds for bone regeneration. *Ceram. Int.* **2017**, *43*, 11265–11273. [CrossRef]

31. Jang, T.H.; Park, S.C.; Yang, H.J.; Kim, J.Y.; Seok, J.H.; Park, U.S.; Choi, C.W.; Lee, S.R.; Han, J. Cryopreservation and its clinical applications. *Integr. Med. Res.* **2017**, *6*, 12–18. [CrossRef] [PubMed]

32. Karlsson, J.O.M.; Toner, M. Long-term storage of tissues by cryopreservation: Critical issues. *Biomaterials* **1996**, *17*, 243–256. [CrossRef]

33. Costa, P.F.; Dias, A.F.; Reis, R.L.; Gomes, M.E. Cryopreservation of cell/scaffold tissue-engineered constructs. *Tissue Eng. Part C Method* **2012**, *18*, 852–858. [CrossRef] [PubMed]

MDPI

St. Alban-Anlage 66

4052 Basel

Switzerland

Tel. +41 61 683 77 34

Fax +41 61 302 89 18

www.mdpi.com

Bioengineering Editorial Office

E-mail: bioengineering@mdpi.com

www.mdpi.com/journal/bioengineering

www.ingramcontent.com/pod-product-compliance
Lightning Source LLC
Chambersburg PA
CBHW051911210326
41597CB00033B/6107